PHYSICS OR METAPHYSICS?

PHYSICS OR METAPHYSICS?

Einstein and Hawking locked in a time-warp – like two flies caught in a spider's web

GERHARD KRAUS

JANUS PUBLISHING COMPANY
London, England

First published in Great Britain 1998
by Janus Publishing Company Limited,
Edinburgh House, 19 Nassau Street,
London W1N 7RE

www.januspublishing.co.uk

A CIP catalogue record for this book
is available from the British Library.

ISBN 1 85756 338 7

Phototypeset in 10½ on 12 Palatino
by Keyboard Services, Luton, Beds

Cover design Nick Eagleton

Printed and bound in Great Britain by
Antony Rowe Ltd, Chippenham, Wiltshire

Contents

Acknowledgment

My thanks are due to Mr Prajuab Nimityongskul, Bangkok, for his valuable help in preparing the manuscript of this book.

Note to Readers

1. All my references to Stephen W. Hawking's *A Brief History of Time* refer to the Bantam Press 1988 edition. My references to Hawking are marked by the letter H, along with the page number; both are bracketed. For example (H32) means: Hawking, page 32. What is important to stress here is that most of my references to Hawking's text are not verbatim quotations but are paraphrased in my own words. Verbatim quotations from Hawking's book are relatively few.
2. To give an authentic view of Einstein's ideas I have chosen Lincoln Barnett's *The Universe of Dr Einstein* (London, Gollancz, 1942) as my main reference source. References are marked by the letter B and the page number, both bracketed.
3. Other important references to Einstein are taken from his work *Relativity: The Special and General Theory* (first edition, 1916). In 1952 Einstein supplemented the text with several editions. This enlarged edition was first published by Mather, London, in 1954. It was reissued in 1960 as a University Paperback. My references are taken from the tenth edition of 1988. They are marked with the letter E and the relevant page number, both bracketed.
4. I draw on the wisdom of Roger Penrose, a close associate of

Hawking, whose *The Emperor's New Mind* was published by Oxford University Press in 1989, and contains many challenging and original ideas on physics. References herein are taken from the Vintage Press edition of 1990. They are marked by the letter P, plus page number, both bracketed.

Other books referred to are named in full.

Preface

In writing this book I have taken recourse to Einstein's own recipes, namely:

a) Never to accept any scientific theory at its face value, if in doubt, and
b) Never to avoid repetitions if this helps textual clarity.

Einstein wrote in his preface to *Relativity* (1916): 'In the interest of clearness, it appeared to me inevitable that I should repeat myself frequently, without paying attention to the elegance of the presentation. I adhere scrupulously to the precept of that brilliant theoretical physicist L. Boltzmann, according to whom matters of elegance ought to be left to the tailor and to the cobbler.'

I cannot of course, rival Einstein's eloquence nor his intuition. Nor do I try to denigrate his truly gigantic contribution to physical science.

The present book is essentially a revised version of my previous book *Has Hawking Erred?* (1992). In this respect it is not only reinforcing my previous argument but offers some radical new ideas. Nevertheless, it also modifies some (though not all) of my previous objections to the concept of space-time.

My questioning of some of Hawking's and Einstein's ideas mainly concerns the following:

1. Aspects of space-time and space-time-curvature.
2. Rejecting the proposition that the practical use of conventional calendar time allows any time dilation.
3. Reviving the possibility of 'absolute-time' and 'absolute-space'.
4. Faulting the principle of a geodesically shaped universe.
5. Doubting (in accord with Einstein) Heisenberg's 'Uncertainty Principle' and, also in accord with him, considering quantum mechanics as a temporary expedient.

It will be shown that the concept of postulating a voluminous space of more than three dimensions and more so that of space-time, where time and space are both subject to bending, slow down and other distortions is *pure metaphysics*.

1

The Role of Commonsense in Physics

Some critics of my preceding work, *Has Hawking Erred?*, have strongly attacked me, by asserting that my ideas contradicting some concepts of Einstein and Hawking are based on plain commonsense considerations. As the following will show, this is not the case, because all my criticisms of the two celebrities are based on commonsense concepts that are guided by logical thinking.

In his book *Superforce* (1984), physicist Paul Davies includes a chapter entitled 'The New Physics and the Collapse of Commonsense'. He maintains that 'the New Physics' defy almost all notions of commonsense, transporting the layman into an Alice in Wonderland world. Davies points out that reality is nevertheless preserved by the wondrous feats of mathematics, which to his mind rule all physical events. At a later stage he goes as far as finding in this physical labyrinth, bolstered by mathematical equations, some striking similarities with the mysticism of Indian and Chinese philosophy. He writes (p. 220):

> Writers such as Fritjof Capra in the *Tao of Physics* and Gary Zukav in *The Dancing Wu Li Masters* have emphasised the unity of existence and the subtle relationship between the whole and its part.

To me, such a total rejection of commonsense pushes modern physics into the realms of metaphysics.

Einstein points to this trend when he says (B12–13):

> In accepting a mathematical description of nature, physicists have been forced to abandon the ordinary world of pure experience, the world of sense perceptions. To understand the significance of this retreat it is necessary to step across the thin line that divides physics from metaphysics. Questions involving the relationship between observer and reality, subject and object, have haunted philosophical thinkers since the dawn of reason.

Hawking (H18) maintains that while commonsense notions may well apply to such earthly objects as apples, and even to our comparatively slow-moving planetary system, they fail to work when we have to deal with objects that move at or near the speed of light.

This is only superficially true, because it can be shown that the notion of 'commonsense' has a very wide-ranging meaning. At least three different aspects of it can be enumerated.

1. *Plain or ordinary commonsense*

This is based on normal sensory experience and its limitations, aptly illuminated in an example cited by Barnett (B14–15), that deals with our faculty of vision. He explains:

> Anyone who has ever thrust a glass prism into a sunbeam and seen the rainbow colours of the solar spectrum

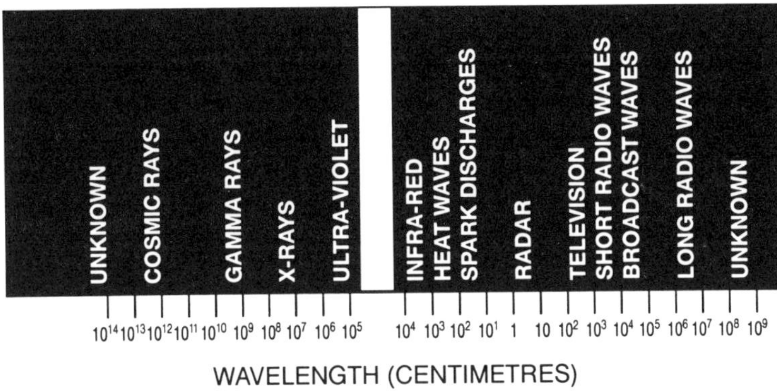

Figure 1: Portion of light visible to the human eye

> refracted on the screen has looked upon the whole range of visible light. For the human eye is sensitive only to a narrow band of radiation that falls between the red and the violet. A difference of a few one hundred-thousandths of centimetres in wavelength makes the difference between visibility and invisibility . . . It is evident, therefore, that the human eye suppresses most of the 'lights' in the world, and what man can perceive of the reality around him is distorted and enfeebled by the limitations of his organ of vision. The world would appear far different to him if his eye were sensitive, for example, to X-rays.

The fact that physicists have been able to overcome the frailty of our inborn biological range of perception is entirely a result of using our commonsense in an ingeniously orchestrated way by devising instruments, leading us to what I call an expansion of commonsense. Thus:

2. *Expanded commonsense*

The range of normal sense perception, enlarged by means of instruments and mathematical calculations. This means that we cannot fathom the intricacies of physical science, which ordinary sense experience fails to unravel, unless we filter them through devices designed by human ingenuity, and elucidated by logical thinking related to scientific facts.

3. *Abandonment of commonsense and logical thinking*

This involves the virtual rejection of ordinary sensory experience, leading to a transformation of physical events into metaphysical phantoms.

I venture to maintain that the science of physics cannot abandon the concept of expanded 'commonsense', for the simple reason that even the most intricate scientific instrument and the most complex mathematical equation is ultimately dependent on normal sensory perception, or what is more obliquely called 'the human mind or intellect'.

The medieval paragon of science, Paracelsus, once said that while God may be able to create an ass with three tails, he cannot create a triangle with more than three sides. Hawking and Einstein, by creating voluminal space of more than three dimensions (including space-time) and going further into five, six, ten and more space dimensions, have created the equivalent of more than three-dimensional triangles. Thus, in the Paracelsian sense, they have outsmarted God.

Sensory perception guided by logical thinking is the ultimate source of all wisdom and forms the fountainhead from which all scientific inquiry emanates, with the proviso that it must be handled with careful consideration and based on ascertainable facts.

2

The Principle of Regularity and Mathematical Vulnerability

One factor in science, the importance of which seems to have hitherto not been given the credit it deserves, is the 'principle of regularity'. Its role, not only in physics but in science as a whole, easily exceeds that of relativity. Or is it perhaps to some degree an alternative term for relativity? Firstly, all measurements of time, space, temperature, weight, sound and gravity, are related to regular units and processes. Secondly, all mathematical formulae, whether relating to abstract quantities or real physical ones, are based on or related to the regularity of numbers. Any law in physical science only gains validity if dealing with quantities, interacting on the basis of regular units of relative measurement. Thus, when Einstein wrote (B38) that 'the laws of nature are the same for all uniformly moving systems', he hit the nail squarely on the head. Regretfully, though, he deviated in his later work from this cast-iron rule when he dealt specifically with the measurement of space and time, adopting the questionable axiom (B46) that 'time and distance are variable quantities'.

Regularity, of course, implies order. Believing in an ordered universe (which is obviously one where the laws of nature interact in a regular manner), Einstein expressed his belief in the famous statement 'God does not play dice'. However, this saying, to strip it of any metaphysical meaning, must be taken in a metaphorical sense and implies the acceptance of regularity. Even the irregularities discovered by Heisenberg in sub-atomic particles become somehow transformed into regularities by incorporating them into the regular equations of quantum mechanics. As to the uncertainty principle of Heisenberg itself, both Einstein and Hawking expressed the opinion that there may be physical laws, as yet undiscovered, which might transform present sub-atomic uncertainties into measurable certainties.

Hawking's comment on this particular aspect of particle physics elucidates that the 'uncertainty principle' as inherent in quantum mechanics deals with events occurring on a sub-atomic level. It postulates that conventional measuring methods fail to pinpoint the position of particles and their velocities simultaneously and with certainty.

However, this apparent unpredictability may only occur when attempts are made to interpret waves in terms of positions and velocities of particles. Hawking suggests that this approach can be mistaken. Perhaps there are neither positions of particles nor velocities, but only waves? And perhaps we try to fit the waves into preconceived ideas about velocities and particles? The resulting misinterpretation, concludes Hawking, may be the cause of an apparent unpredictability.

Using the principle of regularity, events in nature can be synchronised with the decimal system of mathematics. Thus, through having invented a system of numbers which reflects such regularity, humankind has designed a means to unlock some of the most concealed forces of nature. One result was Einstein's famous equation $E = mc^2$, a formula which discloses the awesome powers hidden in the tiny atom.

Similar considerations have persuaded present-day physicists to treat mathematics as an almost infallible medium for revealing the secrets of the physical world. One physicist, Paul Davies (*Superforce*, 1984), invests mathematics with an almost religious awe. The roots of such thinking go back into antiquity. Greek mathematical and astronomical ideas are shown to have derived from Babylonia and Egypt. It is also believed that the Pythagorean theorem was not an original invention of Pythagoras but was rediscovered by him through his contacts with Babylon, where it was known as early as 2000 BC. On the other hand, Herodotus in *The Histories*, reports that Pythagoras learned numbers and measures from the Egyptians. The archaeologist Gordon Childe writes (*Man Makes Himself*, 1966):

> We do not know how any of the geometrical rules were obtained. They certainly had not been deduced *a priori* to be in Euclid's geometry. For such a 'pure geometry' there is no evidence at all. Figures often accompany the geometrical problems in the mathematical papyri and tablets, but such figures are no more drawn to scale than are the fieldplans in contracts. On the other hand, the patterns decorating craft products, piles of bricks and composite wooden boxes often give very striking colour demonstrations of geometrical propositions... Fourteen hundred years later the physical origin of such geometrical abstractions as well as those of numbers and mathematical formulae had become obscure and forgotten. The mystery relationships between squares and circles, between squares and triangles, etc. seem to have been taken over by Pythagoras, as well as Aristotle, Plato and others, as philosophical truths.

Roger Penrose, referring to such traditions (P554), writes: 'According to the Platonic viewpoint mathematical ideas have an existence of their own'.

In other words, all mathematical concepts exist already, concealed within nature's events. The mathematician's task is to discover them; that is, transform physical events into mathematical formulae. Penrose observes (P225):

> It is remarkable that all the SUPERB theories of nature have proved to be extraordinarily fertile as sources of mathematical ideas. There is a deep and beautiful mystery in this fact: that these superb accurate theories are also extraordinarily fruitful simply as mathematics.

And (P554) that 'according to the Platonic viewpoint mathematical ideas have an existence of their own'. Referring to the system of complex numbers, Penrose further remarks: 'With such identification, it might be more comprehensible how "minds" could seem to manifest some mysterious connection between the physical world and Plato's world of mathematics'.

However, human beings are far from perfect, and we do not live in a perfect world. Hence mathematics, although producing phenomenal results, but being a human invention, cannot avoid the pitfalls of human frailty. It is sometimes maintained that mathematical capacity is an inborn property, being one of the latest biologically evolutionary acquirements of modern Homo sapiens, seated in a specific part of the human brain. Evolutionary biology tells otherwise.

Latest studies of human evolutionary history have revealed that Homo sapiens appeared on this earth some time between 400,000 and 300,000 years ago, following the species Homo erectus in the evolutionary scale. What is significant is that ancient Homo sapiens can be shown to have had the same intellectual qualities and potential as contemporary modern man, who has more or less biologically stagnated over the entire period. This means that humans of 300,000 years ago could have theoretically produced the genius of an Einstein or that of any other prominent mathematician, had the opportunity offered itself.

It was not until between 4200 and 4100 BC or so that the invention of writing associated with the use of numbers came into existence in the wake of the invention of a calendar, releasing thereby hitherto dormant mathematical capacities – though initially for purely practical and utilitarian purposes. Gordon Childe writes (*Man Makes Himself*, 1966):

> It must not be supposed that ancient societies were interested in infinite length or empty space. Their abstractions were limited by practical interest. The ancient Sumerian measures of area have in some cases the same names as measures of weight; in particular the smallest unit in both 'tables' is the Se or grain. In other words, Sumerian 'square measure' was originally a seed measure. The Sumerian interest was the quantity of seed needed to sow the fields. He regarded the field not as occupying so much empty space but as needing so much seed grain. With the areas of bits of uncultivatable desert or blue sky he was not concerned.

Theoretical, abstract mathematics with all its modern expressions is a descendant of these rudimentary beginnings.

Despite its phenomenal achievements, modern mathematics cannot claim perfection, being based on the frailties of the decimal system. For example, man has never succeeded in squaring the circle, being unable to produce a conclusive number for pi. Thus in reading 3.14159265358979321846 *ad infinitum* must forever remain inconclusive in mathematical terms, although the relationship it intends to express, namely that of a circle's radius to its surface and to its circumference, is definite and conclusive. The same applies to Planck's constant of quantum mechanics. Further, it can be shown that theoretical mathematics, as projected on paper, does not always coincide with natural events.

Einstein, on occasion, speaks of the stern logic of mathematics (B43), as it is inherent in the simple arithmetical equation of two plus two equalling four ($2 + 2 = 4$). It can be shown, however, in the following, that this is only partially correct, and is not applicable to all events in the natural world. For instance, adding one droplet of water to another will again only give one droplet (though a bigger one), resulting in the relative arithmetical equation of one plus one equalling one (1 + 1 = 1). This purely numerical abstraction pays no heed to either size, form or the nature of objects. Going further and uniting two droplets of water with another two, will again result in only one droplet, the result being 2+2 = 1. By using lumps of plasticine the results will be identical. Rolling two lumps together to make one, the arithmetical formula will again be 1+1 = 1. If we do the same with four lumps, we get the formula 1+1+1+1 = 1, or 4 times 1 is one (4×l = 1). Furthermore, dividing a lump of plasticine into two equal parts we get one minus one equalling two (1–1 = 2).

What these examples teach us is that purely abstract mathematical calculations made on paper, when we try to relate them to practical physical events, do not always match physical realities. This finds confirmation in the mathematics of time-dilation. It will subsequently be shown that the latter's mathematics have no basis in physical reality.

Furthermore, the possibility that certain Einsteinian equations may not necessarily match natural events (which they are supposed to anticipate) was recently demonstrated by two researchers at Cornell University, Stuart Shapiro and Saul Teukolsky, professors of astronomy and physics. They used a supercomputer to simulate a tremendous gravitational collapse in the universe, producing states of matter, an event that – according to Einstein's general theory of relativity – was impossible (see *Physical Review Letter*, 25 February 1991).

It had not been established until then whether *Einstein's equations* could give rise to naked singularities not clothed in black. Dr Shapiro said: '*We don't believe naked singularities actually happen in nature*. So if the theory says that they are occurring that's the breakdown of the theory' [emphasis supplied]. Other scientists who studied the results came to similar conclusions, namely, that the simulation may have exposed a flaw in at least one aspect of Einstein's theory on the behaviour of space, time, matter and gravity.

Another point of the same order concerns extra space

dimensions. Here, attempts abound to prove mathematically (though in doing so only on paper) the existence of a great number of extra space dimensions. Paul Davies (in *Superforce*, 1984, p. 237), cites an example where an eleven-dimensional space feature emerged from lengthy mathematical analysis. Hawking (P162), writing about 'string theories', points out that they seem to be consistent only if space has either ten or twenty-six dimensions instead of the usual four.

As far as Einstein is concerned, we can indicate here in advance of later proof that the same fate of mathematical fallibility enshrouds the Lorentz equations, which play such an important role in his endeavour to establish that time and distance are variable quantities. Einstein asserts (B46) that:

> The scientist who wishes to describe the phenomena of nature in terms that are consistent for all systems throughout the universe must regard measurements of time and distance as variable quantities.

He asserts that:

> The equations comprising the Lorentz transformations do just that. They preserve the velocity of light as a universal constant, but modify all measurements of time and distance according to the velocity of each system of reference.

This, in other words, amounts to a rejection of 'regular calendar time' as the all-embracing medium, which, with the help of clocks or other suitable timepieces, commonly measures the duration of all events in the universe, including the velocity of objects. The problems involved in this proposition are extensively dealt with in Barnett's Chapter 8, *More About Time.*

Finally, to conclude my remarks on mathematical vulnerability, I quote US physicist John Lukacs (1993). Commenting on the proposed giant Texan Supercollider, he writes:

More and more mathematical formulae about subatomic matter consist only of untested and untestable assumptions, all of them theoretical and abstract. The belief that the universe is 'written in the language of mathematics' is not only wrong, it is entirely outdated. 'What is there exact in mathematics except its own exactitude?' Goethe wrote. He was right, as mathematicians in the 20th century have confirmed.

3

Aspects of Relativity

All manifestations in our natural world or, in a wider perspective, in the entire universe, find practical expression only when related to tangible entities. This applies both to organic life and to the inorganic mechanical processes of measurement. In organic life all sensory perceptions relate to sense organs – sight to the eyes, odours to the nose, taste to the mouth, hearing to the ears and skin sensations to the perception of touch, and so on. In the inorganic world all measurements, whether of temperature, weight, sound, density, space (linear and voluminal) and of time, are based on, and related to, arbitrarily fixed frames of reference.

For objects in motion it is necessary to point out that there exists no fixed point in the universe to which measurements of time and position can be related in an absolute fashion. Hence, ever since scientific calculations began, all measurements referring to time, space, temperature, weight, etc., were related to some arbitrarily fixed frame of reference. For this very reason relations between two freely moving systems, although they can be set down in terms of mathematical formulae, are of little practical value and must remain fleeting assessments unless they are related to some extraneous, arbitrarily fixed frame of reference of a third order.

For the latter reason, physicists have generally used our earth or, in a wider aspect, our solar system, as pivots for physical measurements. Thus, all valid human time measurements in our universe are, without exception, based on our conventional terrestrial calendar and its subdivisions; while all linear space measurements are based on our common metric system. Both are founded on criteria which will be outlined later on. In this, our practical application of the measurements of time, distance and volume, no allowance can be made for the slowing down of clocks or the shrinkage of measuring rods, which in the theory of relativity are postulated to occur when clocks or measuring rods are presumed to accompany objects moving at great speeds. As a rule, real, practical measurements are usually arrived at indirectly by calculations on paper. Examples are Hubble's contention that a light ray circling the universe would take 200 billion terrestrial years to accomplish this feat; and that the sunlight takes eight minutes to reach our earth. And although Einstein and his disciples have theoretically proved that measuring rods will shrink and clocks will slow down when accompanying objects travelling at great speeds, this exercise is merely of theoretical value and cannot be seen to have any practical application.

It is noteworthy that most of the above observations on relativity and their interpretation accord with Einstein's own ideas on the subject. For instance, Einstein points out (B40):

> *All the clocks ever used have been geared to our solar system.* What we call an hour is actually a measurement in space – an arc of 15 degrees in the apparent daily rotation of the celestial sphere. And what we call a year is simply a measure of the earth's progression in its orbit around the sun . . . Relativity tells us there is no such thing as a fixed interval of time independent of the system to which it is referred. There is indeed no such thing as 'simultaneity',

> there is no such thing as 'now', independent of a system of reference [emphasis supplied].

Further on (B41) he states:

> It is constantly necessary for the scientist, in dealing with matters involving complex forms of motion (as in celestial mechanics, electrodynamics, etc.), to relate the magnitudes found in one system with those occurring in another.

To this I must add the proviso that one of the two systems (or an arbitrarily designed third system) must serve as a fixed frame of reference. This is again confirmed by Einstein (in *Relativity*, 1988, p. 7): 'If the results of physics and astronomy are to maintain their clearness, the physical meaning of specification of position must always be thought with the above considerations' [i.e. in reference to a fixed system of coordinates; G.K]. Hence Einstein: 'Every description of events in space involves the use of a rigid body to which such events have to be referred'. To this he adds (B33): *'For all ordinary purposes of science, the earth can be regarded as a stationary system'* [emphasis supplied].

While all this conforms with our own views on relativity as set out above, it is remarkable to note (as it will be subsequently shown) that when Einstein deals with the phenomenon of light, he appears to exempt it from the general principles of relativity as shown above, by implying that its proclivities involve a 'deep enigma of nature' (see B43). Einstein elaborates this exemption when dealing with the mathematical laws underlying the relationships known as 'laws of transformation'.

4

The Time Concept

When dealing with the subject of time (as with any other abstract quantity, like space), one must *not* ask – (*a priori*) – what is 'time'? But rather ask (*a posteriori*) – what is the natural phenomenon, or quantity, we describe as 'time'? To solve the problem we must first turn our attention to objects in motion.

According to Einsteinian relativity (which in this particular case matches commonsense experience), there exists no absolute motion in the universe, as all objects move relative to one another. To determine, therefore, and measure a moving object's position in space we must relate it to a stationary frame of reference. Einstein has pointed out that 'for all ordinary purposes of science the earth can be regarded as a stationary system' – thus becoming the pivot, or fixed reference point for any quantitative measuring (be it length, width, height, volume, surface, sound, weight, or whatever). If we then consider any measurable quantity or magnitude as a dimension, then the abstract quantity 'time' can be included in the above list, becoming 'the time dimension'. Up to this point we still have to define what we mean by 'time'.

I believe that the phenomenon we call 'time', specifically

when used in science, only gains an unequivocal, clear-cut meaning when used as a *comparative measure*. Apart from this, it appears that the purely abstract use of the time concept defies a clear definition and somehow strays into the realms of abstract philosophy, like the proposition that physical existence presupposes time by differentiating events in their relative succession. One offshoot of the time concept is 'absolute time', which, seen from a cosmic viewpoint, lacks the prerogative of reality, but which in Hawking's and Einstein's definition (see Chapter 5) inadvertently identifies with conventional calendar time. The seemingly commonsense notion that there is something like an inborn time sense in humans is also questionable.

Penrose (P391) asserts that 'central to our feeling of awareness is the sensation of the progression of time'. This would presuppose the existence of an additional sense-organ with a 'time sense'. And although the succession of events in the universe and their relative progression signifies life, being and existence, as distinct from non-existence (with existence being seemingly time-linked in a philosophical sense), this presumed capacity is not biologically imprinted on the organism. Seen from a biological viewpoint there is no such thing as a sense-organ in either humans or animals which consciously perceives time. The experience of cave explorers who were lost for days in dark caves, cut off from light and sound, and who consequently completely lost any track of the progress of time, has been reported. Equally so, blind people in isolation from sounds and any other sensory experience become completely disorientated as to time perception. Thus, the experience of time seems to be entirely linked to the passing of perceptible events. In practice, it gains reality only when their duration is measured relative to conventional calendar time.

The nearest phenomenon which might be described as a 'time sense' (which it is not), is a reflex action of an organism called the 'circadian clock' or 'rhythm' (Circadian: from Latin 'circa' = approximate, and 'dies' = day.) It is a genetic adaptation of environmental influences such as temperature, climate, the day/night cycle, etc., causing changes in body temperature, blood pressure, the heart beat, hormonal changes and so forth. This biological property is prevalent in humans, animals and plants and cannot be classified as a 'time sense', insofar as it is not a sensory faculty in the conventional meaning of the term, nor is it an instinctive response.

Thus, in its everyday practical use, 'time', like the metric measure, is not more and not less than a comparative means. It is a specific human invention, tailored to compare the intervals between passing events, the speed of moving bodies, the duration of such movements, the ageing of objects both in motion and at rest, as well as the ageing of growing organisms, and overall the age of the universe in which all these events occur, have occurred in the past and, it is presumed, will occur in the future.

Furthermore, all these happenings are attuned and measured relative to our conventional calendar year, itself divisible into months, weeks, days, hours, minutes, seconds and their subdivisions. To carry out time measurements, timepieces have been devised which, to be of any practical use, must work in strict synchronisation with our earth/sun relationship, both for ordinary everyday use and for scientific purposes.

It is important to repeat here that Einstein based time measuring originally on the same premise, asserting that, 'All the clocks ever used by man have been geared to our solar system' (B40).

From a historical viewpoint, time measurement began with the invention of a calendar in pre-dynastic Egypt. It is presumed that it was based on observing the fairly regular arrival of the Nile flood in Lower Egypt. Its prediction became a factor of vital importance for the agricultural pursuits of the Nile Valley dwellers – and thus became a basic element in the origin of Egyptian civilisation. In the Cairo area (before the Aswan Dam was built), the Nile normally rose 23 feet. By the end of September the entire Nile area was submerged and presented the appearance of a large sea of turbid waters.

Settlements built on high ground rose like islands. For a month this water remained stationary, before subsiding rapidly.

The arrival of the Nile flood was always awaited with great anxiety. The archaeologist Gordon Childe has written (*Man Makes Himself*, 1966):

> Now it happens that just about the time the flood should reach Cairo, the last star to appear on the horizon before dawn obscures all stars, was Sirius, the Egyptians' Sothis. The helical rising of Sirius was thus a naturally fixed point in the solar year. It was found that this event recurred roughly every 365 days, and it was then taken as the starting point of an artificial State year of that length. The Egyptian calendar was accordingly divided into 12 months of 30 days each, plus five days extra and an additional day every four years.

Whether a lunar calendar actually preceded the Sothic calendar has never been resolved. On the other hand, it has been suggested (though there is some doubt about the dates) that a more accurate Egyptian calendar was based on the Sothic cycle and that its beginnings can be put at between 4241 and 4240 BC. These dates mark the presumed (though not confirmed) inception of the Egyptian calendar. The strictly astronomical aspect of these dates has been briefly summarized by Bray and Trump in the *Penguin Dictionary of Archaeology*, 1972. They reiterate that the civil and calendar year was based on a year of 365 days rather than the correct 365.422 days, hence lagging steadily behind the true solar year. The length of this cycle was 1,460 years to the time when the two calendars were again exactly in step; this did happen, according to a record of Theon of Alexandria, in AD 139.

As to the modern use of the conventional 'Solar Calendar', the following details are relevant. Two basic kinds of time are now maintained: International Atomic Time and Universal Coordinated Time. International Atomic Time is derived by the International Bureau of Time at the Paris Observatory from three cesium clocks of extreme precision in Canada, West Germany and the United States. Such time runs continuously and is used, for example, to match astronomical observations from different continents or to start up widely separated power-generating systems in phase with one another. The cesium clock at the US Naval Observatory in Washington, like all cesium clocks, churns away the hours at the rate of 9,192,631,770 oscillations per second. Time signals from national services are monitored in Paris, and once a month each service is told if its clocks are running fast or slow.

Universal Coordinated Time, while it runs at the same rate, is designed to keep in phase with the earth's rotation. This is the standard that the everyday world lives by and that governs time announcements on radio or over the telephone. It differs from time as it might be determined by astronomical observations in that such predictable effects as wobbles of seasonal variations in the earth's spin are averaged out. To demonstrate the exclusive use of regular conventional calendar time in all spheres of astronomy and physics, I am citing a number of typical examples, in which I emphasise all references to the regular units of calendar time, such as seconds, minutes, days and years.

Einstein mentions (B41) that the star Arcturus is 38 light *years* away from us; also that the nearest galaxies are estimated at 100 million light *years* away from us and the remotest about 500 million light *years* away. The astronomer Hubble, using Einstein's field equations, has calculated that the radius of the universe measures 35 billion light *years* and that a sunbeam setting out through space at the rate of 186,000 miles per *second*, would encircle the universe and return to its source after a little more than *'200 billion terrestrial years'* (B85/86). Unless the above time measurements are based on a perfectly regular timeflow related retroactively to the earth-sun cycle, they would make no sense.

Hawking (H24) notes that a sunray takes four *years* to reach the star Alpha Centauri; also (H37) that we live in a galaxy which orbits around its centre about once every several hundred million *years*, and further (H46), that between 10 and 20 thousand million *years* ago the distance between neighbouring galaxies must have been zero. He consequently concludes (H108) that the universe is between 10 and 20 billion *years* old. By thus using the principle of 'regular calendar time' and its subdivisions in the above context, Hawking implies that all events in the universe since its suggested beginnings, whether they relate to the large galaxies of stars or the minutest particles of matter, took place within a time framework of about fifteen billion *years*. These *years* are expressed in conventional solar, or calendar, years of a regular time flow, each again counting about 365¼ *days* and further divisible into *conventional hours, minutes, seconds and picoseconds.*

The Big Bang theory of the origin of the universe holds that after the initial explosion of the superhot ball that began it all, a process started that led to the creation of atoms within 500,000 *years* and the early formation of stars and galaxies 200 million *years* later. After about 300,000 *years* from the event, the opaque plasma from the primary explosion turned into more transparent gas, resulting in microwave radiation. All this only makes sense on the assumption that *all the years* mentioned follow each other in perfectly regular sequence. Furthermore, the *years* in question are those of *regular terrestrial calendar years,* which for all practical reasons are based on units, which we can justifiably describe as *regular calendar years.*

Eric J. Lerner, who disputes the 'Big Bang Theory' (1991), points out that 'Because galaxies travel at small fractions of the speed of light, mathematics shows that such large clumps of matter must have taken at least *100 billion years* to come together'; thus, if correct, outdistancing Hawking's assumption of a fifteen billion-year-old universe six times over. While not necessarily accepting Lerner's findings, my point here is again that the *100 billion years in question, if applicable, are conventional terrestrial calendar years.*

A theory postulating that the cosmos is *trillions of years* old furthermore disputes the 'Big Bang'. Again, such a time scale only makes sense when it is assumed that *each of these billions or trillions of years is of equal duration, proceeding throughout the entire period as the same uniform pace.* This uniform progress of what is obviously 'regular calendar time' must prevail, whether the time scale is based on the conventional solar calendar or on any other possible alternative time relationship, attuned to, or synchronized with, conventional calendar time.

It involves the same principle which rules that a distance measured between two points remains unchanged, whether we express it in inches or centimetres, miles or kilometres. For example, Mercury orbits the sun in 88 of our days, and in the same period rotates just once on its axis. So the duration of a Mercury year and a Mercury day amounts to the same thing. A regular Mercury year equals 88 earth days and a regular Mercury hour equals 88 divided by 24, or 3.66 earth days.

Adhering to the same regular calendar time scale, particle physics uses tiny time fractions like *microseconds* and *picoseconds,* the latter being a mere 10^{-12} part of a second of our regular terrestrial time standard. Furthermore, picoseconds, which are a trillionth of a second, are still divisible into femtoseconds. For example, there are more femtoseconds in one second than there were seconds in the past 31 million years.

Referring to the creation of the universe, Penrose, an adherent of the 'Big Bang Theory' points out (P424–5)

> From that moment, *one thousandth of a second* after creation (i.e. 10^{-4}), the emergence of the primordial fireball, until about *three minutes* later, can be described by well-established physical theory ... Thus it was the initial fireball that spread its gas so uniformly throughout space [emphasis supplied].

Again, the seconds and minutes mentioned are retroactively based on conventional calendar time.

Among other phenomena related to conventional calendar time is the *'light second'*. According to Hawking (H23):

> '*A light second* is simply defined as the distance *the light travels in one second (of terrestrial time)*, this being some 300,000 kilometres (186,284 miles). One light metre, again being the 300,000 millionth part of a *light second*, has been accurately measured by cesium clocks and equals in duration 0.0000000033335640952 parts of *a conventional terrestrial second*. Its multiple is convertible *into conventional terrestrial minutes, hours, days and years*, which gain their more accurate expression in "sidereal time". A sidereal day is the period of a complete rotation of the earth upon its axis with respect to the fixed stars, while the sidereal second is the 24th part of a sidereal day, divided by 3600. Furthermore, "regular calendar time", as related to sidereal time, can be synchronised with the speed of light, another absolute constant in the universe'.

Other phenomena of nature which are *strictly related to our conventional terrestrial second* are Einstein's famous formula for energy and the 'Planck's Constant' of quantum theory.

In Einstein's formula $E = mc^2$, E denotes energy, m denotes mass and c stands for the velocity of light which proceeds at a pace of 300,000 kilometres *per terrestrial second* anywhere in the universes.

The German physicist Max Planck discovered that electromagnetic oscillations occur in quanta, where energy E has a definite relation to their frequency v, given by the formula $E = hv$, with 'h' being recognized as being very tiny, equalling 6.6×10^{-34} *'Joule seconds'*. A *'Joule second'* again is a unit of work, equalling 107 ergs, and equals the work done in *one second* by a current of one ampere flowing through a resistance of one ohm. What needs stressing here is that a *'Joule second' is identical in duration to our conventional second*, thus forming a basic ingredient of quantum mechanics.

It is evident that *physicists the world over (including Einstein and Hawking) have measured and expressed all macroscopic and microscopic events in the universe in terms of regular terrestrial calendar time, which accords with the light second (measured in terms of conventional time). Thus they use, as their universal time standard, the main aspect of what Einstein (as shown later) defines as 'absolute time'.*

5

Absolute Time

The term 'absolute' as used in this chapter is strictly limited to what we consider the regular flow of conventional calendar time, inclusive of minor irregularities in the solar system.

Defining the term 'absolute'

In first trying to define the meaning of the word 'absolute', as used in science, I admit that, theoretically viewed, there cannot exist any complete 'absolute' values in our ever-changing universe. The practical use of the term 'absolute' is therefore conditional on the system of reference to which it is related. For example, chemistry has a formula for 'absolute alcohol', which is related to ethyl alcohol of not less than 99% purity. Significantly though, a 100% purity is not conditional. 'Absolute zero', which denotes the total absence of heat energy, is fixed at 273.16 °C below zero. Although this figure is calculated in relation to earthly measurement conditions, which are variable, it is valid throughout the entire universe. In our case, the conditional use of 'absolute time' is based on the regularity of the earth/sun relationship, although, being subject to minor variations, it presents in the long run a conditional 'absolute' as defined above.

As an example of a minor variation in the flow of conventional calendar time, I like to refer to a calculation made by Keven Pang, an astronomer at the US Space Agency's Jet Propulsion Laboratory in Pasadena, who found that in 12,000 BC the earthly day was one 47 thousandth of a second shorter than it is today.

Other examples are the tides, changing weather patterns, ocean currents, polar ice variations – all affect the earth's rotation. The net result is that it is slowing, losing only a few thousandths of a second per century. Some geophysicists have calculated that half a billion years ago the planet travelled 20 hours to make it through the day. In another 200 million years an earth day may last 25 hours.

All this shows that the conditional use of 'absolute time', and I stress the word 'conditional', equating it with the flow of regular conventional calendar time is quite justifiable. However, Hawking and Einstein disagree.

Hawking writes (p. 18):

> Both Aristotle and Newton believed in absolute time. That is, they believed that one could unambiguously measure the interval of time between two events, and that this time would be the same whoever measured it, provided they used a good clock. Time was completely separate from and independent of space. This is what most people would take to be the commonsense view. However, we had to change our ideas about space and time. Although our apparently commonsense notions work well when dealing with things like apples, or planets that travel comparatively slowly, they don't work at all for things moving at or near the speed of light.

Hawking's dismissal of 'absolute time' seems to ignore the

facts, shown in preceding examples, which indicate that all time measurements, both in practical and theoretical physics, are, without exception, based on conventional calendar time, irrespective of whether they deal with the time measurement of macroscopic or microscopic events, and irrespective of whether these events take or took place within the confines of the minuscule space of the atom, or occurred within the universal space of the cosmos. As regards measuring the speed of light, we have shown that it is unequivocally based on the conventional second and the conventional metre. Hawking and Einstein had dismissed the concept of 'absolute time' as outdated, describing it (B39) as *'a steady, unvarying, inexorable, universal time flow, streaming from the infinite past to the infinite future'* [emphasis supplied]. What is remarkable to observe here is that the 'regularity aspect' of time, inherent in Einstein's steady, unvarying, inexorable, universal time flow, is embodied in both Einstein's definition of absolute time, which he dismisses, and in conventional calendar time, which he accepts. He noted (B40) and I repeat. *'All the clocks ever used have been geared to our solar system'*.

Hawking's particular contradiction about the time concept arises from the fact that having, on p. 18 of his book, rejected the regularity aspect of 'absolute time' as defined by Einstein (above), he revives it later in the text, on p. 118, by counting the age of the universe in billions of regular conventional calendar years. This, it must be emphasised, is a method of time measurement *which only makes sense if the years in question follow each other in a steady, invariable and perfectly regular time sequence, streaming from the beginning of the universe into present times – thus equalling Einstein's definition of 'absolute time'*. This shows in other words, that Hawking rejects the regularity aspect of 'absolute time' in theory, yet uses it in practice. Other examples where Hawking routinely uses regular calendar time are his unconditional acceptance of the speed of light, *expressed in regular conventional seconds,* as well as his use of Einstein's formula $E = mc^2$, where c stands for the speed of light, equally *expressed in regular conventional seconds.* Many more examples abound where Hawking and Einstein use ordinary terrestrial calendar time to measure astronomic events in the universe as well as within our solar system, and therefore inadvertently *make conditional use of what they define as 'absolute time'*. They reject this term, yet it is identical with the regular flow of calendar time as defined by Einstein above.

The preceding examples show that all practical time measurements in astronomy and in physics (though not necessarily their theoretical accompaniments), as well as those of everyday life, are based on our conventional terrestrial calendar. It has further been shown that this is based on the universally applicable criteria of the sun/earth relationship. What needs stressing here is that the resulting time readings also apply to events which happened billions of years before our own planetary solar system came into being. More remarkable still is the fact that this time measurement also applies retroactively to the tiniest fractions of seconds, down to a primordial fireball's appearance one thousandth of a second after the creation of the universe. As to macrocosmic events, I refer to Hawking's assumption that the universe is about 15 billion years old. What is notable here is that such an assumption can only hold true when it is also assumed that all the 10 to 20 billion years in question were of equal duration, with each year following another in perfectly regular sequence down to the smallest split second. Hawking's assumption, it would appear, must therefore exclude any possibility of embodying in its system any kind of time irregularity.

I reiterate therefore that time measurement based on our terrestrial calendar presents a system of absolute regularity throughout the history of the universe, disallowing any kind of time slowdown, or even a possible standstill of time as postulated in the time theories of Hawking and Einstein.

Furthermore, the linear distance measurements accompanying all events in the universe enumerated above are based on the unit of our *standard metre,* whose prototype is projected between two markings on a platinum bar lodged in Paris, denoting the metre's length. I am taking astronomer Hubble's postulate that a light ray encircling the universe would take 200 billion years to do so, as an example for metric measurement. To measure the distance the light ray travels during this time we have to translate it into metric units as follows. Firstly, one light year, which is the distance light travels in one year, has to be converted into light seconds. Accordingly, light travelling at 300,000 kilometres per second (299,793 km to be exact) has to be multiplied by $365^1/_4 \times 24 \times 3600$ to get the number of kilometres it travels in one year. The result is 9.4 trillion km (equalling 5.9 trillion miles). This figure is again to be multiplied by 1000 (1000 metres being equivalent to one km). By multiplying the resulting distance again by 200 billion, we obtain the total metric distance the light will have covered during its journey circumscribing the universe, although not necessarily accepting its geodesic spheric shape. I am giving this example only to show that our common metric unit of measurement is uniformly applicable throughout the universe, without allowing any variation in its length, and that it can even be retroactively applied over billions of years back, although its introduction occurred only recently.

These details indicate that our conventional time and distance measures are quite absolute insofar as they are universally applicable and are retroactively valid back to the presumed creation of the universe. This, according to Hawking, occurred c. 15 billion years ago. One may therefore justifiably describe our time and metric measures as expressing absolute values, being reflected in such phrases as 'absolute time' and 'absolute distance'.

Remarkably though, this latter definition of 'absolute time' matches both our own notion of 'absolute time' and that subsequently elucidated by Hawking and Einstein.

6

Can Time Slow Down?

On 23rd April 1992, a meeting of the American Physical Society proclaimed a scientific breakthrough. The Cosmic Background Explorer (COBE) Space Craft had collected evidence, which allegedly confirmed the birth of the universe reaching back about 15 billion years to the Big Bang. It was revealed that between two immensely short time intervals of less than a split second – namely between 10^{-35} seconds and 10^{-33} seconds after the Big Bang, the universe expanded from the size of less than an atom to the size of a grapefruit. In the subsequent expansion of the universe, events were measured in intervals of two minutes; 300,000 years, two billion years, five billion years and fifteen billion years.

In accord with previous conclusions, the time measurements employed were based strictly on the regular flow of conventional calendar time, tuned to the solar calendar. What becomes manifest in this scenario is that these time measurements reach retroactively back to the very birth of the universe, thereby involving events lasting the tiniest part of a split second. All this despite the fact that regular solar calendar time was only introduced a few thousand years ago. What we find here is an example of the practical use of what Einstein (B40) has defined as absolute time, namely *'A steady unvarying inexorable time-flow, streaming from the infinite past to the infinite future'*.

It stands to reason that this definition (which applies fully to the COBE discoveries) does not allow variations in the universal time-flow to occur – much less allowing a total time-standstill, as all events since the universe's beginnings are without exceptions measured in terms of conventional regular calendar time which is unvariable (apart from tiny irregularities which do not affect the overall picture).

Another fact establishing the prominence of conventional calendar time is Einstein's dictum that the speed of light is an absolute dimension. The fact that the velocity of light covering 300,000 kilometres in one second is measured in conventional calendar year seconds, and the distance covered in conventional metres based on the light second, elevates both these dimensions, representing duration, also to absolutes.

In an article published in the *National Geographic Magazine,* titled 'The Enigma of Time' (March 1990), John Boslough records an interview with physicist Stephen Hawking. 'I asked Hawking to explain relative time'. 'There is only one absolute,' Hawking told me. 'It is the speed of light, 186,282 miles per second. In a sense, the speed of light is Einstein's masterclock since there can be no velocity without time and the speed of light is really the best way to measure time.'

Thus Einstein and Hawking confirm that the speed of light is an absolute dimension and thereby inadvertently admit that the conventional second and the conventional metre on which it is based must also be absolute dimensions. In the light of this axiom their assertion that time is a variable entity becomes untenable and contradictory.

I will now cite several examples by Hawking and Einstein where they try to prove and justify that the universal regular time-flow of conventional calendar time can be subject to slow-downs (dilations) and can even come to a total standstill. For instance, both scientists want us to believe that the age of a person is influenced in a sense by the behaviour of clocks they carry. Hence the clock of a person living high above the earth's surface, they assert, will indicate that he ages slower than a person living at sea-level, or that a person travelling at high velocity up into space will slow down in ageing at the same rate as the watch he carries slows down. Hawking (H32) has maintained that these predictions were tested and confirmed by a water-tower experiment in 1962.

According to Hawking (who records the experiment though not being its author) two accurate clocks were used. One was fixed at the bottom and the other on top of the water-tower (whose height was not stated). The result was that the bottom clock, being nearer the earth's surface, was found to run faster than the one at the top and that this happened in accord with general relativity's predictions.

I am now going to show that Hawking's assumptions are questionable by replicating the water-tower experiment, but using sundials for time measuring instead of mechanical clocks. The idea is simple enough to be proved on paper without physically ascending the water-tower.

Assuming the placement of two sundials, one located at the bottom and the other on top of the water-tower, it will be found that at twelve o'clock noon exactly, both dials will read the same time, with a minuscule difference, caused by the fact that the light has to travel a measurable distance between the top and the bottom dial. However this comparative time reading will exactly replicate itself after 24 hours without showing the slightest discrepancy in the time readings. This contradicts Hawking's water-tower conclusions based on the use of mechanical clocks.

Going further, we now assume the presence of two sundials, one positioned on earth exactly synchronized with another fixed on an extended water-tower reaching halfway to the sun. It can be predicted that at exactly twelve o'clock noon earth time (with the shadow of the earth-bound sundial pointing at twelve o'clock noon), the shadow of the sundial positioned halfway up to the sun will register four minutes after twelve o'clock noon, this being the result of the sunlight taking eight minutes to reach the earth, and thus four minutes to reach halfway. After another 24 hours the shadows of the two sundials will point exactly at the same notations as the day before, recording the same time difference without the slightest variation. And this will repeat itself day after day indefinitely.

In the example of the so-called twin-paradox, Hawking suggests that we have a pair of twins, one of them going to live at sea-level and the other on top of a high mountain. When they meet again at sea-level after a considerable time it is found that they differ in their age. The one living at the mountain top would now be younger than the other twin who continued living at sea-level. Hawking has pointed out that the difference of age would be relatively small. On the other hand, if one of the twins had undertaken a long journey on a spacecraft, travelling near the speed of light, the age difference would be considerable. This so-called twin-paradox, comments Hawking, would however remain a paradox only for those who believed in the principle of absolute time.

I like to add here that none of the people who have climbed Mount Everest have ever claimed or experienced a reduction in their age. Neither have astronauts. Several astronauts have visited a space station located 395 miles above sea level and some of them have even stayed there a year or more. Yet it has never been recorded that any one of them experienced or claimed the slightest reduction in his age on his return, in comparison with his compatriots who stayed behind.

The next hypothetical example has been proposed by Einstein. He writes, 'In a Buck Rodgers realm of fantasy, it is possible to imagine some future cosmic explorer boarding an atom-propelled space ship, ranging the void at 167,000 miles per second, and returning to earth after ten terrestrial years, to find himself physically five years younger.'

In my example, I am now proposing to send Einstein's explorer on a second trip, this time to the star Alpha Centauri, a star four light years distant from earth. However, instead of using a space-ship I let him travel piggy-back on a light-beam travelling at the light's speed of 300,000 kilometres per second. What appears obvious is that the round trip (to and fro) by the space traveller will have taken eight years, during which time the light beam (consisting of materially perceptible photons) will have aged eight years, each light year equalling in duration one terrestrial calendar year. I maintain that our explorer riding on the light-beam could not have aged more slowly than the light beam itself, and on his return would have aged exactly eight years in tandem with the light beam.

Now, according to Einstein (B49), 'A clock travelling at the speed of light [presumably a conventional mechanical clock], carried along by the explorer on his second trip, would stop completely.' Therefore, in accordance with Einsteinian relativity, the explorer would also stop ageing. On the other hand, the light beam would still have increased eight years in age because, according to Einstein, the constant speed of light can not be subject to any time retardation. In other words, the light covers a definite distance, during a fixed interval of time, equalling its ageing.

If we then assume that the explorer's age on either of his two trips were timed by a sundial device, his ageing would, on his second journey, coincide exactly with the ageing of the light beam, totalling eight years, while on his first trip (in Einstein's example) he would have aged ten years in accordance with the sundial's reading – coinciding with terrestrial time – not ageing five years as Einstein predicted. Based on these considerations, the theory of the slowing down of time, advanced by Hawking and Einstein, appears to be conflicting with the true progression of natural events. As to the mathematical support of their theory, this may appear quite plausible on paper, though, as will be shown, it still conflicts with the reality of natural events it attempts to illustrate. This ambiguity inherent in some mathematical equations, not always matching nature's manifestations, was dealt with in more detail in a preceding chapter.

In the following I am citing two further examples with

which Hawking and Einstein have attempted to prove their theory of the slowing down of time. In one, Einstein (B51) compares a radiating hydrogen atom to a kind of clock in which the time can slow down. According to him, physicist Ives, who conducted the experiment, compared the light emitted by hydrogen atoms at rest and found that the frequency of vibrations of the moving atoms was reduced in exact accordance with the predictions of Einstein's equations.

What Einstein (in accord with Ives) omits to mention is that, while the frequency in the two cases varied, the speed of light emitted from the same hydrogen atoms, both at rest and moving at high speed, must have remained constant and unchanged, travelling exactly 300,000 kilometres per terrestrial second measured by a cesium clock. As the speed of light is a much more reliable permanent clock than its variable frequency, no slowing down of time can be read into this experiment.

A similar explanation can be accorded to Hawking's example (pp. 32/33) where he attempts to evaluate the 1962 water-tower experiment. He points out that with the introduction of very precise navigation systems, related to satellite signals, the discrepancy in the speed of clocks at different levels above the earth's surface is now of great practical importance. He maintains that if one neglected the predictions of general relativity the result would be a discrepancy of several miles in the calculated position of the satellite under scrutiny.

As in Einstein's example above, relating to comparative frequencies, no slowing down of time can be recorded. What happens is that increased velocities in the frequencies allow an evaluation of the differences of the frequencies at differing distances from the earth, which can aid navigation in outer space. As the speed of light remains thereby unaffected, being a much more reliable time indicator, no slowing down of time is implied.

In drawing wider conclusions about the so-called phenomenon of the variability of time, let us now assume that the light rays from all the galaxies criss-crossing the various regions of the universe do so at speeds expressed in regular terrestrial seconds (a fact fully endorsed by Einstein and Hawking). We cannot accept then that some of the regions they cross can accommodate events which occur at a slower time sequence despite the fact that each region comes under the same terrestrial time standard. In other words, it appears impossible that two differing calendar-time sequences, one fast and the other slow, can co-exist side by side in the same time region of the universe. To assume from this that time based on conventional calendar time can slow down in some regions of the universe, and even in some cases can come to a total standstill, while the light travelling through the same space regions, measured by a sundial clock, retains its regular speed and ageing in terms of terrestrial time is an unacceptable proposition. There can therefore only be one universal time standard, which, as previously exemplified, must be based on the regular flow of conventional calendar time. It is only under such terms that Hawking's estimate of the age of the universe and the age of other events in the universe, as well as Einstein's dictum of the invariability of the speed of light (based on our conventional second), can find practical application.

Let us now reiterate my points about the regularity of our basic time dimension. We have shown that physicists the world over measure all physical events in the universe in units of regular calendar time, and distances in regular metric units based on it. These regular measurements do not permit any variation in the time-flow to occur, nor allow a shortening of distances. However, in the face of such all-pervading universally applicable regularities, Einsteinian relativity maintains that clocks attached to moving objects travelling at high velocities will slow down and measuring rods carried along will shrink in size, thereby influencing the regularity at which objects move.

One can easily refute this Einsteinian dictum of time dilation by means of the following indisputable example. The distance between earth and sun is conventionally measured by reference to a sun ray of light which takes eight minutes to reach our earth. In terms of metric distance this works out at 300,000 kilometres per second (the correct distance being 298,767 km). Hence, by multiplying 8 × 60 300,000, we get a distance between earth and sun being 1 billion and 440 million kilometres.

According to Einstein, a conventional clock moving at the speed of light will come to a total standstill. Equally so, a measuring rod moving at the light's speed will shrink to zero length. Hence, by hypothetically attaching both a clock and a measuring rod to a light ray emanating from the sun towards the earth, they would register zero time and zero distance, thus defying the true time and distance measurements which read eight minutes in time and one billion and 440 million kilometres in distance.

Such reasoning shows that moving clocks and moving measuring rods can in no way influence the real measurements of time and distance of objects in motion and that the assumption that they can do so is pure illusion, having no basis in physical reality. The final conclusion is therefore that time (based on regular conventional calendar time) neither does, nor can, slow down.

7

The Concept of Space

From the earlier part of his life, up to near his demise, Einstein agonised about the meaning of space. At the age of 27, in his book *Relativity* (1916, page 9), he wrote: 'In the first place we entirely shun the vague word "space" of which we must honestly acknowledge we cannot form the slightest conception…'. And still, in 1952, at the age of 73, he expressed similar doubts. In an Appendix to his *Relativity* (page 134), under the title 'Structure of Space According to the General Theory of Relativity', Einstein concluded, 'I further want to remark that the theory of expanded space, together with the empirical data of astronomy, permit no decision about the finite or infinite character of (three-dimensional) space…'. Hence an ultimate conclusion about spatial characteristics still eluded Einstein.

Let us see whether we can add some new ideas to Einstein's dilemma.

In pursuit of this objective I like to say that a re-examination of the space concept reveals several twists in its possible interpretation, some of them being of a semantic nature. Most important is the necessity for a strict differentiation between voluminal and linear space and their relation to time.

Reading through the various contributions of Hawking and Einstein on the subject one finds much ambiguity in their use of spatial facets. Often their distinction between linear and voluminal space becomes blurred, particularly in their treatment of the subject of space-time, and even more so in respect of the curvature of space-time. (Both are dealt with in separate chapters below.) For instance, according to Hawking (H23), time cannot be considered as something independent, or separate from space, but forms, as he asserts, an object termed 'space-time'. He states (H33):

> In the general theory of relativity space and time are dynamic quantities: when a body moves, and a force acts, it affects the curvature of space and time – and in turn the structure of space-time affects the way in which bodies move and forces act.

Hawking does not tell us to which type of space the curvature refers, whether it relates to one-dimensional linear space or to three-dimensional voluminal space.

Dictionaries generally describe 'space' as:

1. The boundless expanse in which objects exist and move.
2. An area or surface of objects.
3. An interval between two points or objects, i.e. space-distance.
4. An interval of space expressed in terms of time.

These varying descriptions, although pointing at the multifarious nature of space, are inadequate for our purpose and need further elaboration. I believe that the following definitions can help to clarify the issues involved.

Firstly, there is one-dimensional linear space which separates two points or objects. It determines the distance between static bodies and the speed or velocity of moving ones. The latter involves the use of time, in addition to linear distance measurement. On the other hand, plain distance measurement between static points or objects such as a line drawn on paper or a railway track laid across country, does not involve any time relationships – as only motion or ageing does.

Secondly, there is two-dimensional area or surface space, which has no material existence of its own. It is determined and measured by the two coordinates, length and width. This two-dimensional space concept is purely abstract, insofar as it expresses a mere attributive property (the surface area), with which all material bodies are endowed, irrespective of their shape or size. Surface space can be identified by its colour or shape or by any other surface particularity which covers or envelops objects, and this applies in a strictly nonmaterial, attributive way. The same principle applies to the surface of a fluid, though with slight differences in perception. Furthermore, surface or area space is a permanent feature of all material objects, whether solid, fluid or gaseous, none of these features having any specific time relationship. As an example I can point to the surface of my table on which I write this sentence. It has no intrinsic time relationship, with one exception – it ages, together with the table, though only hypothetically, having no material existence of its own.

An example follows of how the two-dimensional space concept has been misrepresented by Hawking. It is accompanied by the explanatory illustration of a two-dimensional animal, taken from his chapter 'The Unification of Physics' (H164).

According to Hawking, in the creation of complicated creatures like humans it appears that two space dimensions are not sufficient to sustain them. He supposes that if two-dimensional creatures were to live on a planet that had merely one dimension, in order to pass one another such animals would have to scramble over each other's backs. Furthermore, a two-dimensional animal, when it came to eat, would hardly be able to digest any food. Instead it would have to void the food in the same form in which it had swallowed it, as the digestive passage would cut the animal in two. Therefore, any two-dimensional creature would not only be unable to digest food, but would have fallen apart even before attempting to eat. Similarly, the presence of blood circulation in a two-dimensional being would present a total anomaly.

Such an example presents us with one of Hawking's rare aberrations. To imagine a two-dimensional being it must have, besides length and height, also thickness (which immediately would transform it into a three-dimensional being) even to eat or to scramble over another such being. Even to think of a two-dimensional being is a total anomaly, because the term two-dimensional is only an attributive expression, describing a purely non-material, hypothetical situation. Also, speaking of a one-dimensional world is a total anachronism, because even in the most vivid imagination such a world cannot have any material existence whatsoever.

One must therefore not be surprised to find Hawking, in common with Einstein (as it will be shown later in the text), endowing the abstractions which denote universal voluminal

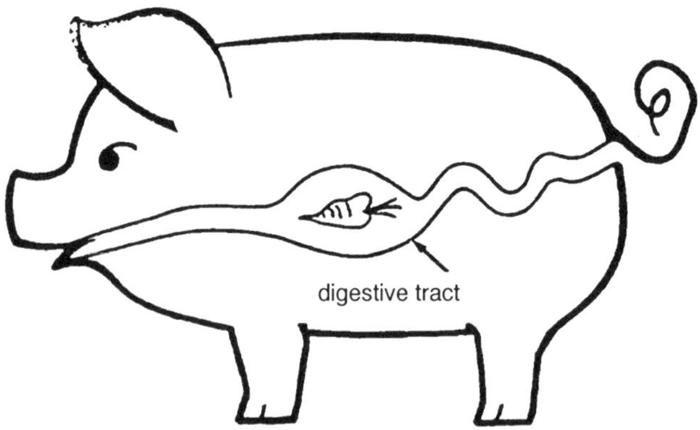

Figure 2: The two-dimensional animal. (Hawking's example is a dog).

space with the capacity of bending, warping, curving and even flattening (like the animal shown), though they have no material existence whatsoever.

Thirdly, there is three-dimensional voluminal space, possibly the most important type to be referred to here. It can be described as an all-pervading, non-material, abstract entity which envelops the entire universe. Being a non-material entity it is completely indifferent to the passage of time. It actually fits Hawking's description of 'absolute space', which he describes as 'a shapeless, steady, hypothetical void, which can be conceived as empty space and which for all practical considerations is static, timeless and of infinite dimension.'

Hawking dismisses 'absolute space' as described by him, above, as obsolete and outdated. And this goes hand in hand with his dismissal of 'absolute time' as defined by Einstein (see Chapter 5, on 'Absolute time'). The reason is, of course, that an acceptance of the principle of 'absolute space' and of 'absolute time' would undermine to some degree one of the premises on which the theory of relativity is built and which demands that 'space' (i.e. 'universal voluminal space') is a malleable substance and that 'time' is an unsteady and variable phenomenon.

To pursue our comments on 'universal voluminal space' further, we can point out that it houses all the three-dimensional voluminal objects and substances extant in the universe and incorporates all one-dimensional configurations, whether time-bound or independent of time, as well as all two-dimensional configurations (i.e. flat surfaces or surfaces enveloping solid objects), which, being merely attributive abstractions, have no time linkage by their very nature as non-material objects.

A remarkable example where a three-dimensional space object (a piece of string) is given a material existence, though it is supposedly a one-dimensional phenomenon, is advanced by Hawking (H159), in commenting on the so-called 'string theories'. Hawking explains that in string theories the essential objects are not particles occupying specific positions in space, but are things that have *only one dimension*, namely length, thereby resembling infinitely thin pieces of string. Strings having ends are called open strings, while strings forming closed loops are called closed strings. In contrast, a specific particle occupies a certain point in space at any particular moment of time, so that its history is traceable as a line in space-time, the so-called 'world line'. A string, on the other hand, at any specific moment of time occupies a line in space which leads to its history in space-time forming a two-dimensional surface called 'world sheet'.

The flaw in this example is the assertion that the strings (intended to represent an alternative to conventional particles) have only length and no other dimension. Now it is quite logical that even the infinitesimally thinnest string, whatever its structure, must have a material existence and be a three-dimensional voluminal space object. Alternatively, a one-dimensional space configuration is purely descriptive and must always remain abstract and nonmaterial. If, in defiance of this principle, it is asserted that the strings have only length, i.e. are one-dimensional, it is clear that they cannot have any material existence whatsoever and in fact cannot exist in a one-dimensional form. Strings in this context embody the same contradiction as Hawking's two-dimensional dog. However, Hawking not only passes over this conceptual error but conceals it within the enigmatic phraseology of space-time, which is intended to rationalise it.

Consideration must now be given to the question of whether universal voluminal space – the space that contains the universe – is basically empty, a question as yet unresolved.

At one time it was believed that the entire universe, including what we term 'empty space', was filled with a hypothetical substance called ether, the medium in which electromagnetic waves, etc., were assumed to be transmitted through space. For example, ether was postulated by James Maxwell to establish the laws of electromagnetism. The concept of ether has since been discarded, but to date no rational analysis has been put forward to explain how light waves and other similar media (including energy) can travel through supposedly empty space without some medium to support them. It must therefore be asked whether an alternative to ether has yet to be postulated, to act as an energy transmitter, filling the entire universe and therefore negating the concept of 'empty space'.

In a more detailed analysis, universal voluminal space reveals itself as a purely hypothetical, intangible entity, timeless and, like surface space, with no material existence of its own. This all-embracing, timeless and spatially limitless void, which we conceive as universal space, houses everything from the biggest galaxies down to the minutest particles of matter. Also included in it is what Einstein describes as a universal gravitational field. Again, all these objects, whether at rest or in motion, occupy part of the volume of universal space, in their own three-dimensional presence.

This definition does not exclude the possibility that universal space may not be entirely empty, but contains, for example, the universal gravitational field postulated by Einstein. However, even if universal voluminal space were entirely full, it still implies the presence of the hypothetical entity of universal space in the first place, so as to accommodate the gravitational field in the second place. The principle involved may be compared with the space in a presumably empty bottle filled with air, except that universal voluminal space must be imagined as occupying a vessel of infinite dimensions. In other words, the hypothetical abstraction 'space' (in the sense of voluminal space) must be present whether it is full or empty.

Einstein himself has pointed to this fact in a footnote on page 155 of his book *Relativity* (1988), in the chapter 'Relativity and the Problem of Space'. He writes, 'If we consider that which fills space (i.e. the field) to be removed, there still remains the metric space'. However, in the very same paragraph, he points out that: 'There is no empty space, i.e. a space without field.'

Penrose, who like Einstein and Hawking unreservedly believes in space-time, comments thus on 'empty space' (P240): 'One implication of Maxwell's equation was that electric and magnetic fields would indeed "push" each other along through *empty space*' [emphasis supplied]. Writing about mass and energy, he argues further (P285) that:

> ...if the latter is to be located at all it must be *in this flat empty space* – a region completely free of matter or fields of any kind. In these curious circumstances, our 'quantity of matter' is now *either there, in the emptiest of empty regions, or is nowhere at all* [emphasis supplied].

On another occasion, Penrose (P436–7) speaks of an astronaut travelling in *'empty space'*, and (P269) states that, '...rays of light (world lines of photons) in *empty space* are also geodesics'. One ought to assume, therefore, that in contrast to Einstein, Penrose accepts the reality of 'empty space', which, to have any theoretical standing, must equal 'absolute space' as defined (though rejected) by Hawking and Einstein.

All this may sound rather involuted. Nevertheless, I believe that the problem of space can be presented in such a way that even the layman can understand its implications. Fundamentally, we can say that 'space' is a descriptive term which, in the course of the development of ordinary language, became incorporated into our vocabulary, allowing us to rationalise it. As already pointed out, the term 'space' itself, like other expressions of multifarious meaning, lends itself to different interpretations, describing a) linear space-distance, b) surface-space, c) volume.

To give a practical example, let us consider an unfurnished, empty rectangular room. The space, or distance between the walls, as well as that between floor and ceiling, can be measured linearly. The surface of the four walls, as well as those of floor and ceiling, can easily be identified as surface spaces, while the hypothetically empty space occupying the room is the room's volume. The fact that the room will normally be filled with air or a gravitational field is of no significance.

We assume now that this room is part of a house separated from another house by a measurable linear space-distance. We also find that the first house has a large backyard and a front garden, while the second house is situated in front of a large playing field. There is no difficulty in describing the backyard, front garden and playing field and other neighbouring open floor spaces as surface spaces, all measurable in two dimensions. Furthermore, the voluminal spaces above and between the houses, as well as those above the backyard, front garden, playing field and any other neighbouring open space, stretch upwards into limitless distance, reminding us of the saying 'the sky is the limit'. Readers will also be acquainted with the term 'stratosphere', which typifies the space above the earth's atmosphere. All the voluminal spaces mentioned above are again part of the three-dimensional universal voluminal space which contains the entire universe. Scientifically considered, universal voluminal space is a purely abstract entity which has no material existence.

If we now also consider that there are material objects located in the open spaces around the houses, such as trees, fences, telegraph poles, stones, etc., there is no difficulty in identifying them as three-dimensional space objects, each of them occupying part of the voluminal empty space around the houses. What is further obvious is that all the above spatial configurations, when applied to the practical needs of everyday life, present no serious conceptual difficulties to persons of average intelligence.

From the above details about space it is further obvious that universal voluminal space, whether empty or occupied, being itself motionless, is independent of the passage of time. On the other hand, any three-dimensional objects when in motion can be time-bound, insofar as their velocities are measurable in relation to regular conventional calendar time.

Let us now assume that I pick up a handy stone and throw it into the open space above me, as far as I can. The stone will describe an arc of 40 metres, reaching the ground again after a flight of, let us say, five seconds. What is obvious here is that the stone will not have lost its three-dimensional qualities while moving through the space above and will land unchanged in volume.

Based on this stone example, I venture to suggest (as further detailed in later chapters) that the same unchanged qualities will prevail if I now extend the example to an aeroplane, a spaceship, the movement of our earth around the sun and around its own axis, the movement of entire galaxies, and the movement of a humble worm, crawling along the ground. None of these objects will be affected in their three dimensionality, whether moving at snail's pace or near the speed of light; although in the latter case (that of high velocity) the shape and mass of the moving object may be subject to change, though not its three-dimensional quality. To assert, however, that moving three-dimensional material objects can gain an additional space dimension is a total illusion. Further, the three-dimensional voluminal space through which all these objects travel will not be affected by the object's passage. Empty voluminal space is a purely hypothetical non-material medium which plays only a passive part in the process and remains totally unaffected by objects passing through it, whatever their speed or composition – although, as Einstein points out, the universal gravitational field will be affected by their passage.

Later in his life Einstein cited examples (see B57–8), which more or less match the descriptions of the threefold nature of 'space' proposed in this book.

For a one-dimensional (linear) space continuum he uses a ruler and a railway track; for a two-dimensional space continuum he uses the sea's surface; for a three-dimensional space continuum he submits the flight of an aeroplane, writing:

> An aeroplane pilot guides his plane through a three-dimensional space continuum, and this constitutes space

as we perceive it. In other words, the space of our world is a three-dimensional continuum.

All this is quite acceptable, except that he fails to mention that this space continuum has no discoverable bounds.

Einstein also tried to make us visualise the non-material nature of space (i.e. of universal voluminal space) by writing (B14): 'Space has no objective reality except as an order or arrangement of objects we perceive in it.' This is a definition which, when everything is considered, implies that space is an abstract, non-material entity matching our own view on space. What is remarkable here is that Einstein's view on space, which he shares with Hawking, is identical with the two scientists' definition of 'absolute space', a concept which both have rejected as obsolete and outdated.

Absolute space

As already indicated, Hawking described 'absolute space' (which he rejects) as 'a shapeless, steady, hypothetical void, which can be conceived as empty space and which is static, timeless and of infinite dimension'. Special reference must be made here to Newton (see B34), who presumed that space itself might serve as a fixed frame of reference to which the wheeling of the stars and galaxies could be related in terms of absolute motion. Newton regarded space as a physical reality, stationary and immovable; while he could not support this conviction by any scientific argument, he nevertheless clung to it on theological grounds, for to Newton space represented the divine omnipresence of God in nature.

Einstein (B38) rejected the Newtonian idea of space (i.e. of 'absolute space'), that is, of a fixed system or framework, absolutely at rest, within which it is possible to distinguish absolute, to which from relative motion, saying that 'in space there are no directions nor boundaries'.

We thus find ourselves with two differing descriptions of

'absolute space', both rejected by Hawking and Einstein as outdated:

a) the Newtonian one, resting on a materially based, perceptible frame of reference;

b) the one put forward by Hawking, above, in which space does not provide a Newtonian-type fixed frame, but which, to repeat Hawking, is definable as a 'shapeless, steady, hypothetical void, which can be conceived of as empty space and which is static, timeless and of infinite dimension'.

I have emphasised the word 'timeless', because it is the separateness of time and voluminal space contained in the above definition of 'absolute space' which forms the essence of their rejection of absolute space as defined by Hawking. This rests on the fact that Einsteinian relativity postulates an intricate relationship between time and space, treating time as an extra fourth space dimension resulting in the concepts of space-time and space-curvature. Einstein writes (B62):

> An understanding of the space-time continuum is requisite to a comprehension of the General Theory of Relativity and of what it says about *gravitation, the unseen force that holds the universe together* and determining its shape and size [emphasis supplied].

The curious result of all this is that the elusive concepts of space-time and space-curvature in association with Einstein's universal gravitational field can be construed as a new fixed frame of reference for astronomical calculations, and thus be considered as a veiled retreat into a Newtonian-like, though not identical, 'absolute space'. This is indicated in the arguments about gravity in my next chapter, because by associating space-time and space-curvature with his universal absolute, according to him (and in retrospect to Newton), the wheeling of the stars and galaxies can be mathematically related.

8

Space-time Analysed

For thousands of years, ever since language bestowed any meaning on these concepts, space (i.e. universal voluminal space) and time have been treated as two distinct natural phenomena, each of them serving a different function. This distinction of function is shared by other natural phenomena such as temperature, sound, weight, and so forth. Just as the functions of temperature and sound do not intermingle, so neither do the functions of space and time, or those of time and temperature: all this in spite of the fact that all events occurring within the realms of universal voluminal space are time bound and temperature bound, while space in this scenario serves as a passive medium.

Furthermore, time and space must always remain different phenomena, and time can never be assumed to function as an extra space dimension, neither in linear nor voluminal space.

Before going into details about the intricate subject of space-time, I find it necessary to recall previous definitions of time and space.

Time

In its everyday practical use, 'time' (like the metric measure) is no more and no less than a comparative means of measure, which has no corporeal existence of its own. It is a specific human invention, tailored to compare and measure the intervals between passing events, the speed of moving bodies, the ages of objects, in motion or at rest, the age of growing organisms and, in a global application, the age or duration of the universe within which all these events occur.

All these occurrences are attuned to our conventional calendar time and its subdivisions. They are based on, and regulated by, the astronomical relations between earth and sun and have been used ever since time measurements attained any degree of accuracy. I recall here again Einstein's remark (B40): *'All clocks ever used have been geared to our solar system'* [emphasis supplied].

Space

In considering the different functions of time and space one should assume as axiomatic that time measurements express duration and linear space measurements distance, and that voluminal space is strictly three-dimensional. This should further imply that linear distance cannot be measured by clocks but only by metric units of length. If one speaks therefore (and this has become common usage) of a) a 'light second' or b) a 'light year', implying a linear distance, this is merely a shorthand expression which actually means a) the linear distance light travels during one ordinary second and b) the linear distance light travels within one calendar year's duration. As to voluminal space, its three-dimensional quality is absolute and cannot be amplified by the addition of further space dimensions. This applies equally to universal space and to all the material objects (all of which are three-dimensional), which it houses, and none of which can accommodate any extra space dimensions, just as a triangle can not accommodate more than three sides.

Curiously though, this cast-iron rule has been breached by Einstein's Special Theory of Relativity, which postulates the existence of the concepts of space-time and space-curvature. My aim is to show that Einstein's postulates are not in accord with natural events, but rest specifically on a turn of phrase involving a wrong interpretation of the term 'space', with the substitution of linear space-distance by time distance. The result has been the conversion of time into an extra space dimension. The distinction between linear space and voluminal space is not strictly adhered to in the writings of Einstein and Hawking. This laxity leads to an intermingling of these two otherwise distinct space phenomena, thereby causing a great deal of conceptual confusion.

Space-time in linear space

The space-time examples given by Einstein and Hawking derive from Euclidean geometry, which in itself is merely a paper exercise; moreover, the examples point to the possibility that the anomaly of an extra space dimension may have arisen from the particular semantic interpretation to which the term 'space' lends itself. One can habitually express 'time' in terms of space, stemming from phrases such as: *a train covers a distance of five kilometres in the 'space' of five minutes*; or light covers a distance of 300,000 kilometres *in the 'space' of one second*. In these and similar examples, 'space' in the sense of linear space (although basically a mere figure of speech, denoting distance) seems identifiable with time (denoting duration).

That Einstein used this particular semantic interpretation when formulating his ideas on time and space is evident in the following passage from his book Relativity (1988, pp. 56–7). Referring to the discovery by Minkowski, his mathematics teacher, that time and space are one, Einstein writes: *'Pure space distance* of two events with respect to K (K being a first coordinate system), *results in time distance* of the same event in respect to K', K' being the second coordinate system)' [emphasis supplied].

As to Hawking, he elucidates (H23) that to fix the position of a specific point in space we define it by the three coordinates, length, width and height. For example, a point within the voluminal space of a room might be situated three metres distant from one wall and five metres distant from another, while its height from the floor measures three metres. In another example (H24), Hawking suggests that one may measure the position of a specific point in the London area in miles north and west of Trafalgar Square; or alternatively, miles northeast and northwest of Trafalgar Square. Equally so, using relativity, one could use a new time co-ordinate, which was the old time (in seconds), plus the distance (in light seconds), north of Trafalgar Square.

In the foregoing we have an example where Hawking expresses a linear space distance in terms of time, elevating time to an extra space dimension. He subsequently concludes: 'It is often helpful to think of the four co-ordinates of an event as specifying its position in *a four-dimensional space called space-time*' [my emphasis].

It would thus appear as if time measurement expressing duration could function as an alternative to metrical measurement expressing distance, thereby elevating time to an extra space dimension. In other words, it appears as if metrical space-distance could be substituted by time-distance (as in the example 'in the space of five minutes'), or even be transformed into time-distance or vice versa. It is possible that the semantic principle expressed in this context, although applicable here only to linear space, may have induced Einstein to generalise (the more so as the German expression 'Zeitraum' is more explicit), and along with other considerations add time as a fourth space dimension to three-dimensional voluminal space.

Hawking also writes (p. 23): 'In the theory of relativity, *we now define distance in terms of time and the speed of light'* [emphasis supplied]. He also refers to the *light-second* as a convenient new unit of length measurement, and writes: *'This is simply defined as the distance the light travels in one second.'* [Emphasis supplied].

Hawking further points out that as space is not completely separate from time, they unite to form the concept of space-time. Just as there is no particular differentiation in relativity between time and space coordinates, so there is no real differentiation between two space coordinates.

My argument is that Hawking's conclusions drawn from the above account and other similar examples are highly questionable. Firstly, the fixed standard distance of a metre (although now expressed in fractions of a light second) was in use long before anybody thought of using light as a comparative measuring rod. Furthermore, using the awkward fraction of a light-second for the measurement of distances in everyday life seems quite cumbersome and is indeed impractical. On the other hand, the use of the light-second, or the light-year, for measuring astronomical distances is quite practicable.

Yet even so, it is quite illusory to assume that the velocity of light denoting distance can serve as a surrogate for a clock denoting duration. The reason is that a light-second or a light-year is only casually related to distance, because what it really denotes is duration. To define these terms correctly, one has to say that *a light-second denotes the distance light travels in one second's duration and a light-year denotes the distance the light travels during one year*. What is most important to stress here is that the second and the year in question equal in duration our conventional second and our conventional calendar year – as a matter of fact, they are based on it. Hence, during the same one second's duration in which the light covers a space distance of 300,000 kilometres, an athlete may cover a space distance of ten metres on a running track, the light-second being exactly the same as any ordinary second. Or while, over a light-year, light travels a distance of 365 × 24 × 36000 ×

300,000 kilometres, the duration of time involved equals exactly that of a conventional calendar year. The stem of a tree may grow merely ten centimetres taller during the same light- or calendar year, the light-year equalling exactly, and being identical to any calendar year.

In the statement that light covers a distance of 300,000 kilometres in the 'space' of one conventional second, which is of the same duration as a light-second, it is clear that the term 'space' stands here for 'duration', measurable on any conventional clock or stop-watch.

To illuminate the principle of space-time, Einstein provides the graphic example of a railway journey, shown below.

The space-time diagram of Einstein (B59)

A train leaves New York at 6 a.m. (Eastern Standard Time), reaching Chicago at 1 p.m. (Central Standard Time) the next day. The intermediate stations are marked with the times noted vertically and the space-distances horizontally. According to Einstein the diagram illustrates a space-time continuum, elevating time to an extra space dimension.

Einstein explains:

> To describe any physical event involving motion, it is not enough simply to indicate position in space. It is necessary to state also how position changes in time. Thus to give an accurate picture of the operation of the New York–Chicago express, one must mention not only that it goes from New York to Albany to Syracuse to Cleveland to Toledo to Chicago, but also the times at which it touches each of those points. This can be done either by means of a time table or a visual chart. If the miles between New York and Chicago are plotted horizontally on a piece of ruled paper and the hours and minutes are plotted vertically, then a diagonal line properly drawn across the page illustrates the progress of the train in a

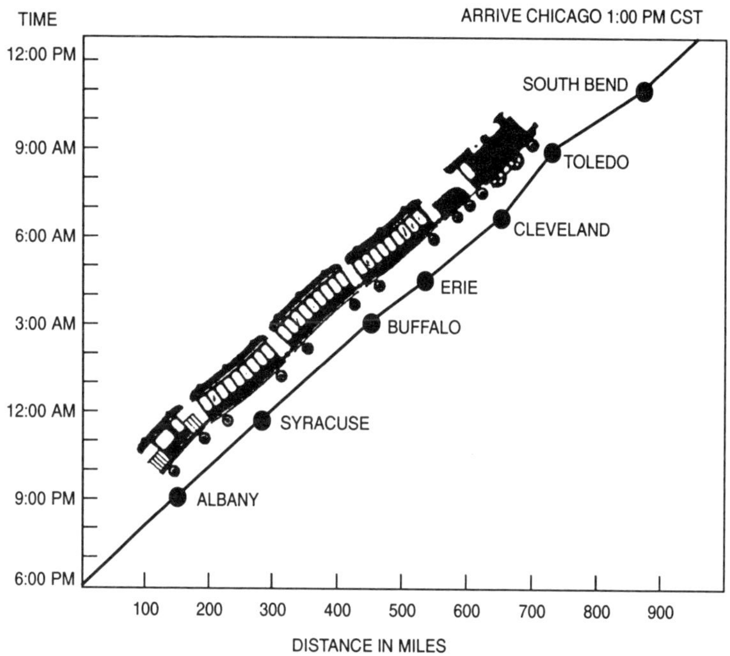

Figure 3: The westbound run of a New York–Chicago express pictured in a two-dimensional space-time continuum

> two-dimensional space-time continuum. This type of graphic representation is familiar to most newspaper readers: *a stock market chart, for example, pictures financial events in a two dimensional dollar-time continuum* [emphasis supplied].

On closer examination this example reveals various flaws. For instance, the train is bound to stop at all the stations listed for five to ten minutes or even longer. This means that during these stopping times, totalling perhaps one hour, the clock, or clocks, recording the progress of the train will keep on ticking, even though the train is at a standstill, covering no distance in terms of linear space at all during the intervals. This shows that Einstein's train example, which is supposed to show continuous motion (illustrating a continuum) in terms of linear space, synchronised with time-duration, and thereby illustrating space-time (a unified concept), is misconceived. What it shows rather is that while the space-distance involved as related to time is discontinuous and studded with interruptions, the related time duration shown on a clock is continuous and uninterrupted. In other words space-distance here does not match time duration: the assumed space-continuum differs from the time-continuum.

Thus, contrary to assumptions, it shows that the New York–Chicago journey, with the train covering 1100 miles in the time-space of 19 hours, does not synchronise with the progress of time during the stoppages. Hence, although Hawking's original reference to linear space-time (H23) – where he writes: 'We must accept that time is not completely separate from and independent of space, but is combined with it to form an object called space-time' – seems to indicate that time, superficially seen, can play the part of an extra space dimension, closer examination reveals that this is not the case.

Einstein's train example makes this quite obvious, so that we can now add that the space (or time-space) in question does not directly relate to the linear distance the New York–Chicago express travels, but to the linear distance (or space) covered by the dial of the Chicago station clock, which indicates the distance the hand moves along its circular face.

The time-space, or distance traversed by the hand on the face of any conventional clock, would, however, shrink to nil if a digital or quartz clock were used. Furthermore, these time measurements in terms of space registered on the face of a clock, whether recorded circularly or digitally, occur quite independently of the linear distance (or space) the train travels during the time in question, or the time-space measured in terms of linear space on the clock's face, traversed by its hands. As the station clocks of New York and Chicago are doubtless synchronised, the clocks would also record at the same time the increase in the age of all the passengers travelling on the train, which would exactly parallel the age increase of the train itself. What is also worth noting is that the age increase of the train and that of its passengers has nothing to do with the space or distance the train travels, as the age increase would remain the same whether it was measured in relation to a stationary train or to one in motion.

These train examples endorse previous conclusions that time and space belong to different categories of phenomena, just as do time, sound, temperature and weight. It can therefore also be contended that time, being a man-made comparative measuring device, recorded by clocks or other timepieces, cannot have any direct bearing on the space factor.

'Space' is an inherent part of nature, while time, or a clock recording it, is a man-made tool arbitrarily aligned to the sun/earth relationship. Just as a metre-rod lacks the inherent potency to regulate the distances between objects or to influence their velocities, and a thermometer lacks the capacity to influence temperature or weather, so does a clock, or other time-measuring tools, lack the capacity to influence or determine events happening in space.

Based on such considerations, Einstein's assertion (B60) that there exists an equivalence of space and time can no longer be upheld; nor can we support Einstein, when he says (B60):

> It must not be thought that the space-time continuum is simply a mathematical construction. The world is a space-time continuum; all reality exists both in space and time, and the two are indivisible. *All measurements of time are really measurements in space, and conversely measurements in space depend on measurements of time* [emphasis supplied].

The dictum that measurements in space are really measurements of time is also contradicted by the following examples. In accordance with accepted theory, the sun's light travelling at a velocity of 300,000 kilometres per second will take eight minutes to reach our earth. To show further the independence of time-measurement (i.e. duration) from space-measurement (i.e. measurement of distance or volume), let us consider what happens within the space (i.e. duration) of the same eight minutes in different parts of the universe. In any region of universal space, light will travel a distance of 8 × 60 × 300,000 kilometres. During the same eight minutes a snail will somewhere on earth crawl a distance of about two metres. A solid rock will merely increase eight minutes in age without showing any measurable space relationship. Equally, any human or animal, whether alive or dead, will increase eight minutes in age, again without any measurable relationship to space. And finally, the entire universe will age exactly eight minutes in perfect synchronisation with my own wristwatch or any other timepiece wherever located on earth. This also means that my wristwatch will indirectly record at the same time the duration of all the events happening in the universe, including the movement of the galaxies and the ages of all objects whether in motion or at rest, with some of these showing, and others not showing, any measurable space-relations. Furthermore, all these events occurring in the universe will take place within the expanse of universal voluminal space, which is a permanent and static void, and is completely indifferent to the time-measurement registered on my watch or any other timepiece wherever located. In other words, my wristwatch or any other timepiece, all being man-made tools for time-measurement, can on no account influence events in the universe, whether in relation to objects in motion or objects at rest. The same principle is inherent in a thermometer. Being a mere tool for temperature measurement, a thermometer cannot exercise any influence whatsoever on temperature or the weather.

Time as an Extra Space-dimension

If we elevate 'time' to the status of an extra space dimension, as Einstein and Hawking would like us to do and as Einstein tries to achieve in his New York-Chicago express train diagram, we can do the same thing with weight, temperature and sound, adding them as extra dimensions to space.

How this can be accomplished is indicated by Einstein himself. Having (in B58) graphically illustrated the progress of the New York–Chicago express in a two-dimensional space-time continuum, he then writes:

> This type of graphic representation is familiar to most newspaper readers; *a stock market chart, for example, pictures financial events in a two-dimensional dollar-time continuum* [emphasis supplied].

This example shows that Einstein elevates time to an extra dollar dimension, although how time, recorded on a clock mounted on the wall of the New York Stock Exchange, can influence stock market quotations is debatable. However, to compare the two-dimensional dollar-time continuum with a two-dimensional space-time continuum is a rather fantastic feat, but no more fantastic than the mixing up of the two entirely incompatible natural phenomena of space and time and fusing them together into the new concept of space-time.

This incongruity becomes even more obvious when we consider Einstein's and Hawking's dictum that in a reverse process to time being a dimension of space, space can become a dimension of time. As the same principle must apply to the dollar-time relationship, we can now say that a dollar can become a dimension of time. Not only that, but having shown that the dollar can function as a time-dimension, while time can function as an extra space-dimension, it must (using Einstein's reasoning) also be theoretically possible to consider the dollar as an extra space-dimension.

However one may look at such two-dimensional manipulations, I maintain that in using Einstein's graphical illustration of a two-dimensional space-time continuum, it is perfectly legitimate to use the same principle of graphic illustration to elevate the natural phenomena of weight, temperature and

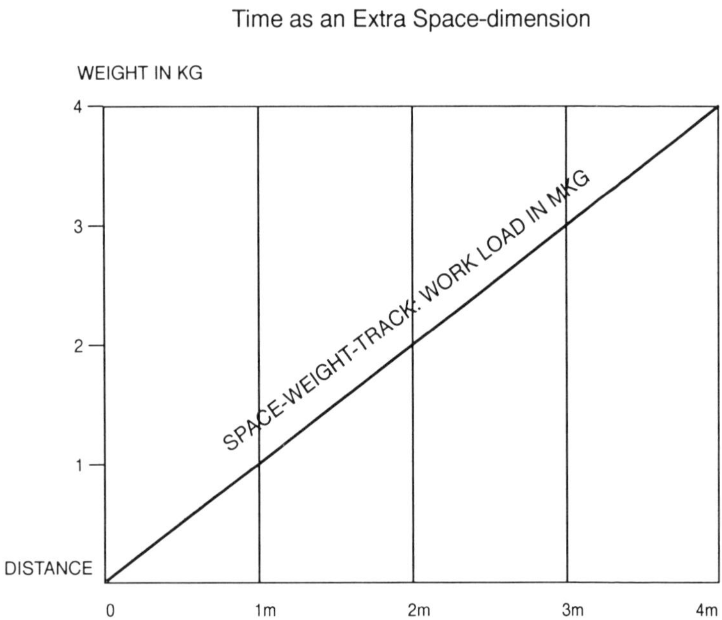

Figure 4: Space-weight diagram

sound to the status of extra space- dimensions. This is illustrated in Figure 4.

The practical unit of work done by a force of one kilogram of weight over a distance of one metre equals one metre kilogram (mkg); that of 2 kg of weight over a distance of two metres will amount to 4 mkg; 3 to 3 will give 9 mkg; and 4 to 4 will give 16 mkg. The weight in kg is marked on the vertical line and the distances the weight moves are marked in metres of linear space on the horizontal line. The diagonal line indicates the respective points corresponding to weight and distance (space). Based on Einstein/Hawking reasoning, we can speak here of *a space-weight continuum, elevating weight to the status of an extra space-dimension.*

In Figure 5, a siren at point A sounds an alarm to the police

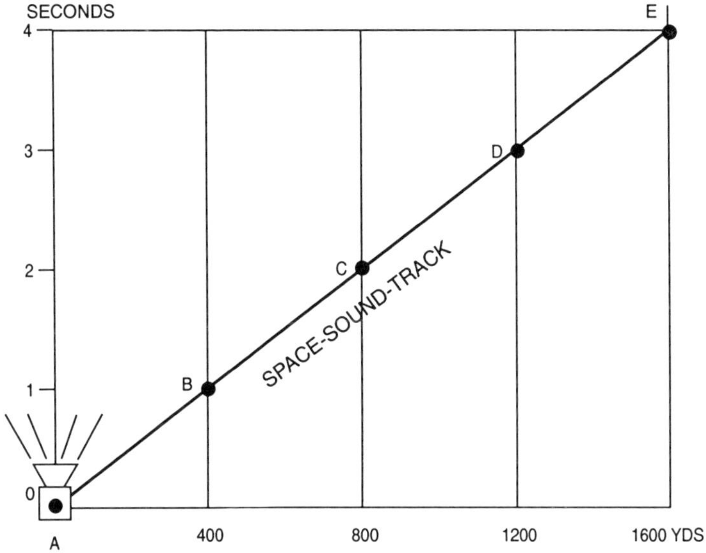

Figure 5: Space-sound diagram

posts B, C, D and E, which are marked on the diagonal line in distances 400 yards apart; the sound, travelling at about 400 yards per second (the actual figure being 373 yards), will reach B in one second, C in two, D in three, and E in four seconds. The seconds are marked vertically and the sound-distances horizontally. The diagonal indicates the position of each respective police post and the corresponding times and distances in which the sound will reach it. Based on Einstein/ Hawking reasoning, we can speak here of a *space-sound continuum (or even a space-time-sound continuum), elevating sound to the status of an extra space-dimension.*

Space – Temperature Diagram *(see Figure 6)*
The sun rises at point A at 4 a.m., when the thermometer indicates a temperature of 10° Celsius. At 6 a.m. the temperature will have risen to 15°, at 8 a.m. to 20°, at 10 a.m. to 25° and at midday to 30°C. The Celsius degrees are marked on the

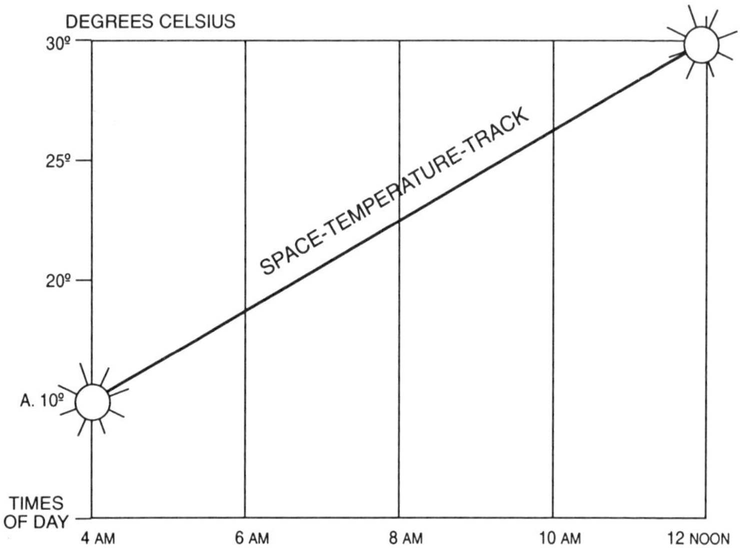

Figure 6: Space-temperature diagram

vertical line in units of height, corresponding with the mercury column of the thermometer. The times of day are marked horizontally. The diagonal line combines temperatures with the respective sun positions. Based on Einstein/Hawking reasoning, we can speak here of *a space-temperature continuum (or even of a space-time-temperature continuum), elevating temperature to the status of an extra space-dimension.*

What the comparative diagrams attempt to show is that, when closely examined, neither time, sound, temperature or weight can be treated as extra space-dimensions. They must rather be considered as figures of speech only casually associated with space. Thus the expression 'in the space of five minutes' can be compared with such phrases as 'deep thought', 'being short of breath', or 'a broad smile'; none of them has anything to do with space measurement. Equally, 'a hot pursuit' has nothing to do with temperature and 'a heavy heart' has nothing to do with weight. Further, the phrases 'a soft heart' or a 'weak response' have nothing to do with physical hardness or weakness. A more extreme example is perhaps 'born with a silver spoon in his mouth', which has nothing to do with physical birth, but implies rather inherited wealth. Similarly, the phrases 'it took a long time for him to grasp the problem', or 'he waited only a short time', have nothing to do with metric distances. 'Long' and 'short' here strictly denote time duration recordable by clocks.

Comparing thus the actual diagrams in this manner, the phrases 'in the space of five minutes' or 'it took a long time' are not directly related to the measurement of a distance in linear space; nor does 'a rise in temperature' measured in degrees Celsius and marked in units of length along a thermometer imply any permanent space-temperature-relationship. The sound of an alarm bell alerting a distant police post does not imply a permanent sound-space-time relationship, nor that of weight to distance. Thus none of the above phenomena can play the part of an extra space-dimension, as all of them belong to different categories of natural phenomena which do not mix. It is the same principle which does not allow an intermingling of sound and temperature, temperature and weight, and weight and sound.

9

Space-time in Voluminal Space

Universal voluminal space is a natural phenomenon, non-material and of limitless dimensions, within which all material objects coexist. The space distances between these objects, whether they are in motion or at rest, are measurable in linear metric units. All material objects again are three-dimensional and possess volume. In dictionary terms, volume is defined as the amount of space (often expressed in cubic units) that a three-dimensional object occupies; in other words, three-dimensionality stands for volume and volume for three-dimensionality.

This three-dimensional quality of all objects in the world is axiomatic, insofar as there cannot exist any material object with either more nor fewer than three dimensions. And this three-dimensionality is just as absolute as the three-sidedness of a triangle. And just as a triangle can never break out of its three-sided frame, so can material objects never escape from their three-dimensional prison, nor squeeze in additional space dimensions.

Now, while it is impossible to prove mathematically, even on paper, that a triangle can have more than three sides, the theory of relativity claims to have proved mathematically (though solely on paper) the existence of a four-dimensional space-time continuum, implying thereby that material objects, normally confined to three-dimensionality, can have more than three space-dimensions. This postulate violates one of the most fundamental laws of nature, thus showing that there are occasions when mathematics can lose its compatibility with physical reality. (For more details on this, see the preceding Chapter 2, Mathematical Vulnerability).

Yet this possibility was inadvertently admitted by Einstein when he wrote (B59):

> Since time is an impalpable quantity, it is not possible to draw a picture or construct a model of a four-dimensional space time continuum. But it can be imagined and *it can be presented mathematically.*

Hawking (P24) writes: 'It is impossible to imagine a four-dimensional space called space-time. I personally find it hard to visualize three-dimensional space.' Yet surprisingly, and against his own better judgement, he presents us with several four-dimensional space-time diagrams (pp. 26 and 27 of his book). The result is doubtful, because all of them appear incomprehensible; although most readers, struck with awe at such profound wisdom, may uncritically accept them at their face value, or else will pretend comprehension.

One need not, however, be deeply versed in modern physics or mathematics to form a sane judgement about an extra, fourth space-dimension within voluminal space, because in physical reality, accepted without prejudice, it is absolutely axiomatic that voluminal space or space objects ('voluminal' simply being another term for 'three-dimensional') can have only three dimensions.

While Euclid could certainly not have intended to postulate a thing called space-time, he demonstrated more than two millennia ago, though solely in writing, that one can generate space of four dimensions mathematically; furthermore, there is no need to stop at four. In extended Euclidean geometry we can go on to spaces of five, six, seven or more dimensions. In developing the general theory of relativity, Einstein found it necessary to adopt a four-dimensional general Riemannian geometry. Instead of a fourth, strictly geometrical space-dimension, Einstein made time his fourth dimension. Einstein pointed out that there is nothing mysterious in this concept. It merely means that every event that takes place in the universe is an event occurring in a four-dimensional world of space-time.

Although Minkowski and Einstein based their propositions originally on linear space projections (as illustrated by Einstein's example of the New York–Chicago express train) both Minkowski and Einstein subsequently transferred this linear space-time concept into the realm of three-dimensional voluminal space. This purely speculative transfer (an obviously metaphysical exercise) from one space medium into another (i.e. linear into voluminal) was remarkable insofar as it subsequently involved the adoption of any number of additional space-dimensions, the transfer of linear space-time into voluminal space and the curving of voluminal space and the treating of universal voluminal space as if it were a plastic medium.

What Einstein also points out (E52) is that:

> The inherited prejudices about time and space did not allow any doubt to arise as to the prime importance of the Galilean transformation for changing over from one body of reference to another. Now assuming that the Maxwell-Lorentz equations hold for a reference body K, we then find that they do not hold for the reference body K' moving uniformly with respect to K, if we assume that the relations of the Galilean transformation exist between the co-ordinates of K and K' ... (E56) As a consequence pure *'space-distance'* of two events with respect to K, *results in time-distance* of the same events with respect to K'. But Minkowski's discovery, which was of importance

> for the formal development of the theory of relativity, does not lie here. It is to be found rather in the fact of his recognition that the four-dimensional space-time continuum of the theory of relativity, in its most essential formal properties, shows a pronounced relationship to the three-dimensional continuum of Euclidean geometrical space [emphasis supplied].

Also, in Einstein's other example given below, we find that the space coordinates he mentions can only refer to linear space-distance, the measurement of which is independent of time. Having once established the position of a body with the help of these coordinates, however, a time relationship can then be established with another body in motion, the position of which is also determined by linear coordinates.

To explain the system of coordinates (E 5–8) Einstein writes:

> Every description of the scene of an event or of the position of an object in space is based on the specification of the point on a rigid body (body of reference) with which that event or object coincides. This applies not only to scientific description, but also to everyday life.

Einstein's example is Trafalgar Square in London. He writes:

> If I analyse the place specification, 'Trafalgar Square, London', I arrive at the following result. The earth is the rigid body to which the specification of place refers; 'Trafalgar Square, London', is a well-defined point, to which a name has been assigned, and with which the event coincides in space.

There then follows a cloud example:

> If, for instance, a cloud is hovering over Trafalgar Square, we can determine its position relative to the surface of the earth, erecting a pole perpendicularly on

> the Square, so that it reaches the cloud. The length of the pole measured with the standard measuring-rod, combined with the specification of the position of the foot of the pole, supplies us with a complete place specification. On the basis of this illustration, we are able to see the manner in which a refinement of the conception of position has been developed.
>
> (a) We imagine the rigid body, to which the place specification is referred, supplemented in such a manner that the object whose position we require is reached by the completed rigid body.
>
> (b) In locating the position of the object, we make use of a number (here the length of the pole measured with the measuring-rod) instead of designated points of reference.
>
> (c) We speak of the height of the cloud even when the pole which reaches the cloud has not been erected. By means of optical observations of the cloud from different positions on the ground, and taking into account the properties of the propagation of light, we determine the length of the pole we should have required in order to reach the cloud.

Einstein continues:

> From this consideration we see that it will be advantageous if, in the description of position, it should be possible by means of numerical measures to make ourselves independent of the existence of marked positions (possessing names) on the rigid body of reference. In the physics of measurement this is attained by the application of the Cartesian system of co-ordinates.
>
> This consists of three plane surfaces perpendicular to each other and rigidly attached to a rigid body. Referred to a system of co-ordinates, the scene of any event will be determined (for the main part) by the specification of the lengths of the three perpendiculars or co-ordinates (x, y, z) which can be determined by a series of manipulations

> with rigid measuring-rods performed according to the rules and methods laid down by Euclidean geometry.
>
> In practice, the rigid surfaces which constitute the system of co-ordinates are generally not available; furthermore, the magnitudes of the co-ordinates are not actually determined by constructions with rigid rods, but by indirect means. If the results of physics and astronomy are to maintain their clearness, the physical meaning of specifications of position must always be sought in accordance with the above considerations.
>
> We thus obtain the following result: every description of events in space involves the use of a rigid body to which such events have to be referred. The resulting relationship takes for granted that the laws of Euclidean geometry hold for 'distances', the 'distance' being represented physically by means of the convention of two marks on a rigid body.

So far, Einstein.

What must immediately come to the reader's mind is, firstly, that Trafalgar Square, London, cannot count as a well defined 'point', but is a large surface which is at places of irregular height and width. Secondly, to use a floating cloud (because no cloud can be described as static) as a specific event above Trafalgar Square which can be located (symbolically) by an exactly measured rigid pole, is a rather questionable decision. A cloud of any description is an irregular vaporous entity that is bound to change its shape or volume practically from moment to moment. To measure its height (or rather that of its underside) seems an altogether precarious venture. One has to conclude therefore that the cloud example is ill-chosen.

However, despite the questionable choice of Einstein's space-coordinate example, certain facts are apparent: all measurements of his space-coordinates are related to linear space-distances; and no time relationships are attached to any of the coordinates, in spite of the fact that Einstein categorically maintains that all measurements in space depend on measurements of time (see B60). It is only when the relative position of different systems in motion is considered that the time factor becomes involved.

Generally, Einstein's examples given above have to be viewed with the following facts in mind:

1. All material objects in the universe are three-dimensional, including the objects whose position is specified in Einstein's coordinate systems of K and K'.
2. To determine the position of any material object in voluminal space, linear space measurement in the form of metric units of length must be employed. This is of course also obligatory when measuring the distances of each of the three space coordinates.
3. None of the linear distance measurements so far cited shows the need for any time relationship, nor does the process of measuring and determining the position on the object within voluminal space (including K and K') influence the dimensional qualities of the objects measured, insofar as their three-dimensionality remains unimpaired. Neither is the empty space (although it may be filled with air, ether or a gravitational field), within which the measurement is carried out, affected in any way by the process of the measurement itself. The example of a football kicked into the space above a football field may serve as an illustration. The football, being a three-dimensional space object, will not only not change its three-dimensional nature while flying through the space above, but will also leave no impact on the voluminal space through which it travels – the latter playing a totally passive role in the process.
4. So far the time factor has not been mentioned. Let us now assume that we want to determine the position of the football when reaching its highest point above the football ground while travelling through the air. Let us assume that this happens at a special point (i.e. Einstein's

point of event), exactly ten metres above the ground at one hundredth of a second after four o'clock in the afternoon on a specific date. What now transpires is as follows.

The timing of the event, as can be observed, makes absolutely no impact on the three linear space coordinates that served to measure the event and determine its position in space; neither does the timing of the event have any impact on the curved flight path of the football itself, nor on its three dimensionality. Furthermore, if we measure the linear distance of the flight path which the football describes while being off the ground, we cannot discover any change in the one-dimensionality of measurement, which is strictly linear. Only if we measure the duration of the time which the football spends in the air before reaching the ground does the time factor come into play. The question then is, how does the time factor influence any particular aspect of the entire transaction from the moment the football is kicked into the air up to the time when it lands back on the ground (including the determination of its highest point of flight)?

The answer is that the time factor exercises no influence whatsoever on the football's spatial aspects except in recording the duration and possibly the speed of its flight. This recording of the duration of the flight on a clock is an entirely passive act, following the same principle which generally applies to the measurements of linear distances or of temperatures. Neither the metre-rod used in the distance measurement of the flight path of the body, nor the thermometer used in the temperature measurement of an object, leaves any impact on the objects measured. As to the time recording the duration of the motion of an object as it covers a linear distance, this exercises no influence on the space dimensionality of the object timed, and more specifically does not change its voluminal status. Time measurement, as shown previously, is simply a process of recording time duration measured by a timepiece. It has no other function.

10

The Enigma of the Curvature of Space

It is difficult to find in the writings of Einstein and Hawking a rational explanation of what they exactly mean when they speak of 'the curvature of space'. We find no clear clue to its how and why. When it emerges, its sudden entry is like that of a stranger from outer space.

Already the plain phrase 'curvature of space' (treating space like a material object) is quite enigmatic, because although both authors have disposed of ether, which would give voluminal space at least some semblance of reality (equalling ether with empty space), they somehow associate space-curvature with gravity. When further linked with the time factor, as when they speak of the curvature of time and space, the phrase becomes even more mysterious; although in the explanation of space-curvature that follows, the time factor seems to play no essential part.

The two scientists surprise us by suddenly bursting on the scene with such statements as (Einstein, B29):

> Gravity is not a force like other forces, but a consequence of the fact that *space-time* is *not flat*, as has previously been assumed, *but is 'curved' or 'warped'* by the distribution of mass and energy in it [emphasis supplied]

while Hawking (p. 33) points out that:

> before 1915, it was natural to think that space and time went on forever. Thus bodies moved, forces attracted and repelled, but time and space went on forever. The situation is now different. In the general theory of relativity space and time are now dynamic quantities; when a body moves, or a force acts, *it affects the curvature of space and time* [emphasis supplied].

Neither man tells us, however, how 'the curvature of space' is achieved nor what the phrase actually means. Do they mean the curvature of linear space, a principle which shows at least some linkage to reality, or the curvature of voluminal space, which appears to be totally irrational?

To trace the origin of such notions, one has to refer back to Minkowski, who can be considered the originator of the space-time principle. In this respect I have to reiterate Minkowski's dictum that space and time have to be considered as a single entity – a four-dimensional space-time – leading to the categorical statement that only a union of the two, that of space and time, will preserve an independent reality.

This, in my opinion, purely speculative and rather philosophical assertion, unaccompanied as it is by any valid explanation of how the fusion of space and time is actually accomplished, and of what Minkowski exactly means when he speaks of 'space' and 'time' (insofar as he fails to define these terms), was uncritically accepted by Einstein and subsequently by the physics establishment including, of course, Hawking. It became a categorical imperative, leading to the further assertion that space and time could bend, warp, curve and even flatten as if they were material substances.

To point to the absurdity of such assumptions I will recall some facts about space and time which are relevant here.

Firstly, linear one-dimensional space and two-dimensional surface space can only gain reality when applied to the measuring of actual objects. Without such an association they are merely descriptive phrases without any physical presence, expressing solely attributive properties. Thus, unless a metric measure is used to measure the distance (whether straight or curved) between objects, 'one-dimensional space' must remain an abstract phrase. And the same holds for two-dimensional space.

Pure abstract 'space' in any of its three forms – one-dimensional linear, two-dimensional surface and even three-dimensional voluminal – are descriptive expressions which denote the measurement of linear distance, surface and volume, all three of them being merely abstractions. They cannot be treated like material objects and therefore cannot be subject to any physical manipulation. They cannot bend, warp or flatten, nor can they register temperature, weight or gravitational effects. Only material space objects, whether solid, fluid or gaseous, can be subject to physical influences. On the other hand, all material objects in the universe are three-dimensional space objects and can therefore be subjected to bending or other forms of spatial distortion.

Let us now reconsider the abstract entity of universal voluminal space. It houses all the objects that exist in the universe, from the biggest galaxies down to the tiniest subatomic particles of matter, all of these being three-dimensional voluminal space objects. What role does one-dimensional linear space, whether straight or curved, play in this arrangement? The answer is (and I repeat): it is only a descriptive phrase which denotes a distance between material objects; or if objects are in motion, it denotes the distance they travel. However, such measurements only gain validity when related to a fixed frame of reference. Thus, the abstraction 'linear space' only gains meaning when used in association with material objects. Without such an association it is nothing and remains a purely semantic word construction.

Furthermore, all distance measurements occur within the bounds of universal voluminal space, which, as we have shown, is the hypothetical void which houses the universe. In this scenario, universal voluminal space plays the part of a passive medium, just as the voluminal space over a football field, through which a football is repeatedly kicked, remains unaffected by the play going on within its bounds.

The abstract term 'space' only gains reality in reference to material things. And as all material objects which exist in the universe, whether solid, fluid or gaseous, are three-dimensional voluminal space objects, they can all (in contrast to the former abstractions) be subjected to spatial distortions like bending, warping, flattening, shrinkage and expansion, and to the effects of temperature, weight or gravity. Unless it can therefore be shown that the term 'space' standing on its own can have a material existence – and the three purely abstractive space concepts, one-dimensional, two-dimensional and three-dimensional, do not have – the assertion that they can be subject to spatial distortion, in their abstract state is, and must remain, unanswerable.

Furthermore, in their abstract, non-material form, none of the three spatial concepts has any specific time relationship – just as a line drawn on a piece of paper (presenting a one-dimensional linear space-example) is indifferent to time. Neither has the voluminal space above a football field have any specific time relationship, irrespective of whether or not a game of football is going on within its bounds or not.

The role of time vis-à-vis space

In Chapter 4, I showed that the purely abstract term 'time' lacks any meaning unless associated with the measurement of the time interval of physical events in motion, or the age determination of material objects. However, such time measurements only gain practical use, both in everyday life and in science, if related to our conventional terrestrial calendar, which again is strictly attuned to the regular astronomical movement of the earth around its own axis and to its encirclement of the sun.

It is remarkable that, although our terrestrial calendar is only a few thousand years old, it is retroactively applicable to events which date back to the assumed creation of the universe. This has been estimated to have occurred between ten and twenty billion years ago, an estimate that has now possibly been pushed further back by some physicists to a hundred billion years or more.

There is even now a theory that postulates that the cosmos is trillions of years old. An article in *Time* magazine (2 September 1991), entitled 'The Big Bang Under Fire', suggests that 'a few scientists believe it is the Big Bang theory itself that is wrong. Instead of a universe that exploded into being 20 billion years ago and grew by way of gravity's tug, they postulate a cosmos trillions of years old and shaped not by gravity but by electricity and magnetism'.

Not only do these billions and even trillions of years, if applicable, represent a regular flow of calendar years all of equal duration, but they retroactively embody time divisions of the tiny fractions of seconds. Here I recall the Big Bang (the creation of the universe) theory, which proclaims (P424–5) that one thousandth of a second after creation there emerged a primordial fireball which can be described by a well-established physical theory. What is important to mention here is that these primordial parts of a second are identical with the fractions of our own conventional second.

Even more remarkable in this reference is that the regular flow of calendar years (or even that of split seconds), being an arbitrary, man-made, comparative measuring device, goes on relentlessly and pays no heed whatsoever to events, which simultaneously happen throughout universal voluminal space. Equally, the earth/sun cycle on which our time standard is based is paralleled by earthbound clocks, which again, while recording the regular time flow, pay no heed to events happening in the universe. In this context, any time relationship to events is only casual, serving specific human needs for solving practical problems.

What must now be obvious to all is that time in the above context, being a purely abstract, comparative medium which denotes a regular time flow, has no material existence of its own and can therefore not be subject to any physical manipulation. Yet Einstein and Hawking, while considering time as an ingredient of space-time, subject it to bending and other physical distortions, assert that time can bend and that space, being equivalent to time, must therefore also be subject to bending. Thus, in space-time, both time and space are treated as if they were physical substances. This further assumes indirectly that the earth/sun relationship is subject to bending, and that on a smaller scale the space in my room can slow down, or in a contrary situation can speed up. That such theories do not match reality is incontestable. The notion that time can bend and voluminal space can slow down can only be described as fantasy.

Now, while the slowing of space and the bending of time defy physical reality, I am quite sure that there are skilful mathematicians who by means of some complicated equations are able to prove that the above actions are theoretically possible. As a matter of fact, such equations are part and parcel of the space-time theory.

My endeavour to learn how the concept of the curvature of space originated is outlined below, though I admit I may have missed the gist of the problem. First, to repeat the homily by Einstein (B57) cited on page 110:

> A non-mathematician is seized by a mysterious shuddering when he hears of 'four-dimensional' things, by a feeling not unlike that awakened by thoughts of the occult. And yet there is no more commonplace statement than that the world in which we live is a four-dimensional space-time continuum ... Once the meaning of the word 'continuum' is properly grasped, the picture of the universe as a four-dimensional space continuum – and

this is the view which underlies all modern conceptions of the universe – becomes perfectly clear.

Or does it?

According to Einstein, a ruler or a railway track are examples of one-dimensional space continuums. But as we have pointed out before, both are static entities which are independent of time. And although Einstein defines a flat surface as a two-dimensional continuum and voluminal space as a three-dimensional one, no true continuity is demonstrable in either, nor can one discover any link with what is called the 'curvature of space'. The problem becomes even more protracted when we move from a one-dimensional static situation to a volatile one, involving the movement of objects.

It is quite obvious that a train and a railway track (both being separate entities) progress in a continuous line. Actually, the train line is only hypothetically continuous, because in real life the train leaves no continuous trace in the space through which it travels. Equally so, a ship moving on the surface of the sea, although also travelling hypothetically in a continuous line, leaves no trace in the water, which, immediately the ship has passed, closes up behind, resuming its former static condition. If we come to the flight of an aeroplane moving through voluminal space (elaborately described by Einstein as a three-dimensional space-time continuum, with time added as a fourth space-dimension), we find exactly the same occurrence as in the case of a train or ship. The aeroplane describes a continuous line, which is merely imaginary and will leave no trace in the space through which it travels.

At this point Einstein and Hawking enter the field, complicating the story by bringing in gravity. An aeroplane, although apparently following a straight path, will, when flying parallel to the earth's surface over a long distance, describe a curved path, which, induced by the earth's gravitational pull, follows the round shape of our globe (a so-called geodesic). Worthy of note here is that the gravitational pull will be on the aeroplane, not on the curved path the plane will take, which delineates a purely imaginary curve and leaves no trace in space, just as a football kicked into the air will leave no trace in the space over the football field.

However, Einstein and Hawking insist that the curved path the aeroplane describes is a real physically perceptible curve, as if the plane was carrying a spaghetti-making machine, leaving a long continuously curved trail of spaghetti behind. This rather drastic example of the curvature of linear space simulates the idea of the two scientists, presenting the space curve of the aeroplane as if it were a solid substance.

So far we have dealt with examples wherein the so-called space continuum (whether straight or curved) is merely imaginary, and does not form a truly continuous line. The next two examples deal with what appear to be genuinely continuous space entities. The first is a lightbeam passing the sun (see Figure 7), the second a jet of water directed towards a burning house.

According to accepted theory, light moves both in the form

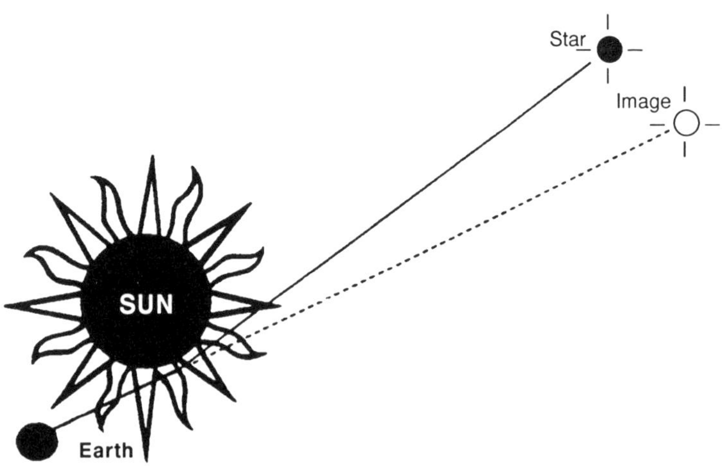

Figure 7: Lightbeam passing the Sun

of waves and in beams of particles which have mass. Einstein accordingly conjectured that light must be subject to gravitational pull. Thus the light of a distant star when passing a massive body like the sun could, induced by the sun's gravitational pull, be deflected from its straight path and form a bend. To test the phenomenon, one had to avoid the glare of the sun and arrange the test during a solar eclipse. When Einstein's assumption was tested under such conditions in 1919, his prediction proved correct. (See Figure 7.)

However, no such lightbeam coming from outer space can be a permanent feature, as it is bound to shift or disappear from the earth's view as the earth moves along in space. As Hawking rightly remarks (H32):

> As the earth orbits around the sun, different stars appear to move behind the sun and have their light deflected. They therefore change their apparent position relative to other stars.

The bending of light rays, which theory proves to possess mass and therefore to be capable of being deflected by gravity, can be simulated (at least in principle) by the example of a water jet directed towards a building on fire. Like the mass of the lightbeam, the mass of the water jet describes a path curved by the earth's gravity. Both examples, lightbeam and water jet, present us with a continuous line, passing through what for all practical purposes can be described as empty voluminal space. When both cease to flow, neither lightbeam nor water jet will have left the slightest trace in the space through which they have passed.

Of further interest here is that a lightbeam, besides being subject to bending, also forms a continuous line. Expressed in Einsteinian terms, it is a linear, one-dimensional space continuum, which with the time factor added as a bonus, makes it a space-time continuum; although in all our previously submitted models of space continuums, such as a ruler, a rail track, a ship, an aeroplane, a lightbeam and a water jet, time did not intrude on, or intermingle with, the space factor, which belongs to a different category of phenomena. In the timing of events, time merely plays the part of an independent, comparative measure.

Referring further to the curvature of space and its credibility, Hawking (H31) turns the light example inside out, by writing that 'the fact that space is curved means that light no longer appears to travel in straight lines in space'; thereby making the bending of the light seemingly responsible for the curvature of space. The use of such arguments in explaining the curvature of space is further emphasised when Hawking (H33) describes time and space as dynamic entities. When forces act and bodies move, they affect the curvature of time and space. In contrast, the structure of space-time influences the movement of bodies and the action of forces. Time and space not only influence but are themselves influenced by all the events that occur in the universe.

In further pursuit of this principle, quite fantastic space-configurations were born. This finds expression in Hawking's book (H135). Commenting on the 'Origin and Fate of the Universe', Hawking refers to Einstein's idea that curved space-time is represented by the gravitational field, whereby particles try to follow the nearest thing to a straight path in space that is curved. However, as space-time cannot be characterised as flat, they describe a bent path as if being influenced by a gravitational field. With all respect for Hawking's other remarkable achievements, such arguments seem forced and convoluted. Yet Hawking uses this enigmatic phraseology to elevate space-time and its curvature to the status of a universal phenomenon.

The most remarkable omission in Einstein's and Hawking's reasoning is their failure to explain how the principles of space-time and of the curvature of space become applicable to three-, four- (and more) dimensional voluminal space. All the examples they advance in this respect are squarely based on linear space projections (a ship, an aeroplane, a lightbeam or a water jet progress strictly linearly). Yet both scientists, suddenly and without further explanation, transfer the principle of linear space into voluminal space, although there appears nowhere any justification for this surprising transposition.

That Hawking has adopted the basic principle of *linear space* to explain space-time and subsequently applies it to both voluminal space and the curvature of space-time, finds confirmation in his book:

Page 22: Under the space-time diagram, Fig. 2.1, he writes, 'Time is measured vertically ... the observer's path through space and time is shown as a *"vertical line"'* [emphasis supplied].

Page 24: 'In relativity there is no real distinction between space and time co-ordinates'.

Page 25: In Fig. 2.2, another space-time diagram, all the space coordinates are again given in terms of *linear space.*

Page 30: 'In general relativity, *bodies always follow straight lines* in four-dimensional space-time but nevertheless, appear to move along curved paths in our three-dimensional space' [emphasis supplied].

Page 31: 'The general theory of relativity predicts that the light cones of points near the sun would be slightly bent inwards by gravitational fields'. [*My comment: light cones thus projected proceed strictly linearly.*]

Page 33: 'In the general theory of relativity space and time are dynamic quantities: when a body moves, or a force acts, it affects the curvature of space and time ... (to which I must remark that this space and time principle is *only applicable in relation to linear space*)' [emphasis supplied].

The same objections hold true in Einstein's case, where he explains space-time by means of linear space examples. There is a ship and a train example on B42, two further railway examples on B44 and B45, and others as well, *all of them strictly linear.*

The most important of these is Einstein's aeroplane example (B58), which attempts to simulate the space-time concept. He writes:

> The flight of an airplane from New York to Los Angeles can best be pictured in a four-dimensional space-time continuum. The fact that the plane is at latitude x, longitude y and altitude z means nothing to the traffic manager of the airline unless the time co-ordinate is also given. *So time is the fourth dimension* [emphasis supplied].

Let us analyse is example once more, firstly by referring to Einstein (also B58), who writes:

> An aviator guides his plane through a three-dimensional space continuum, hence he has to consider not only longitude and latitude, but also his height above the ground. The continuum of an airplane pilot constitutes space as we perceive it. In other words, the space of our world is a three-dimensional continuum.

What is obvious here is that when Einstein speaks of our earthly space being a 'three-dimensional space continuum', he defines a concept which equals our own definition of 'universal voluminal space'.

Also obvious from Einstein's example is that the three space coordinates, longitude, latitude and height, are all one-dimensional linear space projections, by means of which the position of the aeroplane is determined. It can be assumed that all the measurements are made in conventional metric units of length. These measurements, all of them linear one-dimensional, are completely indifferent to the universal voluminal space within whose bounds their measurements are carried out; the universal voluminal space remains unaffected by the act of measuring and plays an entirely passive part in the transaction. As to time, which is supposed to form an extra, fourth space-dimension, the following principle applies.

As already pointed out, space and time are two distinct phenomena, with space denoting distance and time denoting duration. Time cannot therefore play the part of an extra space-dimension, since the regular time flow recorded by timepieces (and related to the earth/sun cycle) is only casually related to the flight of the aeroplane. For example, the same clock or watch which monitors the position of this particular aeroplane records at the same time the minutes it takes to bring a pot of tea to the boil, as well as recording the pot's own ageing. In the two latter cases no space factor is involved. In the case of the tea, to use the Einstein rationale, one could well conjecture that time plays the part of an extra 'temperature dimension'; while in the case of the timepiece recording its ageing, the time factor remains entirely wrapped up in its own time-dimension.

In the light of such examples, which could be extended a hundredfold, Hawking's assertion (H23) that time is not separate from space, but combines with it to form an object called space-time, loses all credibility. Nor can we accept the Einstein assertion (also quoted by Hawking (H23)) that 'In relativity there is no real distinction between space and time co-ordinates, just as there is no real distinction between space co-ordinates'.

We have seen that Hawking's and Einstein's space-time examples, which, besides pointing to the obsolescence of space-time, also relate to the curvature of space and time, are all squarely based on linear space projections, although both authors identify them subsequently with the properties of an assumed four-dimensional voluminal space continuum. Yet in accordance with what was demonstrated above, the principle of linear space in relation to time measurement cannot be imposed on three-dimensional voluminal space. As a non-material void it can neither bend nor warp, nor can it be in any way affected by any object passing through it, be it an aeroplane, a light cone, a jet of water or a football. None of these or any other object can leave the slightest trace in empty voluminal space, either during or after its passage. The postulation of a fourth voluminal space-time dimension and its curvature, postulated apart from linear space, is therefore quite incompatible with the criteria of voluminal space as defined in this work. In other words it is a metaphysical fantasy.

To sum up this chapter I would like to say this. Both Einstein and Hawking bestow space-time with the properties of bending, warping, flattening and slowing down, meaning that each, time and space separately considered, must also be subject to the same influences, namely that time can bend, warp, curve and flatten and that space can slow down. This logically includes the space in my room (in which I write this) which must also be subject to slowing down and flattening.

Furthermore, if the non-material abstractions of space and time can be subjected to the above physical acrobatics, then other non-material abstractions, like temperature, and even hunger, thirst, freedom, hope, fortune, spirit, soul, excellence, etc, etc, must also be subject to slowing down and bending, however absurd this may seem.

Seen in such a context, the fantastic concept of space-time *being blatant metaphysics* can be considered as a major aberration in the history of science, leaving thus a stain on the reputation of both Einstein and Hawking. Yet, while thus tainting them, I like to stress that Einstein and Hawking have contributed major achievements to the physical sciences – which will remain a lasting sign of their genius.

11

Velocity of Light and the Solar Calendar

An interesting fact is the association of the speed of light with our solar calendar. This was already briefly referred to in Chapter 10 in relation to 'Train example 3' and is linked to Einstein's particular interpretation of the Michelson-Morley experiment.

Barnett comments (B37):

> The Michelson-Morley experiment confronted scientists with an embarrassing alternative. On the one hand, they could scrap the ether theory which had explained so many things about electricity, magnetism and light. Or, if they insisted on retaining the ether, they had to abandon the still more venerable Copernican theory that the earth is in motion.

Einstein's assessment of it reads (B38): 'The one indisputable fact established by the Michelson-Morley experiment was that *the velocity of light is unaffected by the motion of the earth'* [emphasis supplied].

According to Barnett, Einstein conjectured further that:

> the velocity of light is constant regardless of the motion of the sun, moon, star, meteor or other systems moving anywhere in the universe. From this, Einstein drew his broader generalization, asserting that the laws of nature are the same for all uniform moving systems.

Einstein's failing here is based on his unconditional acceptance of the dictum *the velocity of light is unaffected by the motion of the earth.* The ambiguity in this statement rests on the consideration that the velocity of light, or rather its measurement, while not directly affected by the motion of the earth, is nevertheless theoretically linked to the motion of the earth.

This thesis rests on the contention endorsed by Einstein himself that the speed of light is measured at 186,284 miles per 'second'. The 'second' referred to is our own earthly second, being part of our earthly time standard. Its duration relates to the fraction of the time the earth needs to move once around its own axis in one day. The entire day again totals 86,400 terrestrial seconds. In other words, the measuring of the velocity of light rests on our earthly time system, identical with what we have defined as 'regular calendar time'. The earthly linkage to the speed of light is further emphasised by the fact that the distance the light travels in one second, i.e. 186,284 miles (or about 300,000 km), is based on our earthly metric system (expressed above in the equivalent of miles), which is again based on our earthly standard metre. The latter's length is synchronised to fractions of an earthly second, accurately measured by cesium clocks.

These examples show that the speed of light is, in the first place, determined or measured relative to the motion of the earth. However, if we abandon the relativity principle and use the velocity of light as an independent absolute constant, unrelated to the earth or any other heavenly body, then calendar time must also become a universal constant. In other words, the postulate of the absolute speed of light is intrinsically associated with that of 'regular calendar time', elevating the latter to another universal absolute in terms of which the duration of all events in the universe can be measured. In other words, conventional calendar time can be *conditionally* put on par with 'absolute time' as previously defined by Einstein – though rejected by both Einstein and Hawking. It was Einstein's persistent rejection of our terrestrial time-standard, itself identical with 'regular calendar time', which became a constant irritant in his own endeavour to reconcile what he describes as the seemingly irreconcilable conflict between his belief both in the constancy of the velocity of light, and in the principle of the addition of velocities; although, as he points out, 'the latter appears to *rest on the stern logic of mathematics, i.e. that two plus two always make four'* [emphasis supplied]. Einstein has described this simple equation as a fundamental law of nature (B43).

This apparent conflict between the two concepts prevalent in mathematics raises doubts that the use of mathematics in physics is not always reconcilable with nature's manifestations. Physics' unconditional reliability on mathematics can by no means be taken for granted (see Chapter 3).

12

The Origin of the Universe

A discovery made in 1992 seems to confirm the view long held by some, though by no means all, of the world's leading physicists that the 'Big Bang theory' so far offers the best explanation of how the universe may have come into being. The details were announced at a meeting of the American Physical Society, held in Washington, DC, on 23 April 1992. They seemed to confirm previous assumptions that about 15 billion years ago a tiny ball of concentrated mass (with no identifiable source) exploded in a 'big bang' and in expanding created the universe as we find it today. The discovery was made by the Cosmic Background Explorer – COBE – spacecraft of the USA National Aeronautics and Space Administration, whose Space Flight Center's computer recorded blotches of all sizes, indicating a region in the universe where the background radiation microwaves are a minuscule 30 millionths of a degree warmer than average. This almost imperceptible variation in temperature, it is asserted, is the missing link now confirming the Big Bang theory.

Also registered were wispy wrinkles in giant clouds of

microwave background, which are estimated to have appeared about 300,000 years after the Big Bang. Earlier instruments, less accurate, had shown that the microwave background was incredibly smooth, suggesting that particles could not have clumped together to form clusters of great galaxies. Hawking, who had long predicted the possibilities now revealed by COBE, hailed this as the scientific event of the century. Carlos Frenk, a physicist at Durham University (Reuters, London, 25.4.92), commented further:

> The discovery confirms scientists were correct in looking for another missing link in explaining the universe's dark matter. The matter that scientists have been able to find – stars, galaxies and other formations – would now explain why there is so much observable gravity. Gravity, the force that keeps people from flying off the surface of the earth and which keeps the earth in orbit around the sun, is produced by matter.

The scenario suggested by the findings of the COBE spacecraft is summarised in the following five points:

1. The Big Bang theory states that all matter was concentrated in a tiny pinpoint, smaller than the finest speck of dust, infinitely dense and infinitely hot, until it was released in a mighty cosmic explosion 15 billion years ago.
2. During what is called the inflationary period, the universe between an age of 10^{-35} of a second and 10^{-33} of a second, expanded from the size of less than an atom to the size of a grapefruit.
3. Within a period of between three minutes after the Big Bang to 300,000 years after the Big Bang, the universe expanded into a fog of free-moving charged particles. During the end of it the expanding cosmos cooled and the particles combined into atoms. Some areas were denser than others, forming gargantuan gas clouds.

4. In the period of between 300,000 years and 2 billion years after the Big Bang, under the influence of gravity, the colossal clouds gradually broke up into smaller, galaxy-sized structures.
5. Between 2 billion years ago and the present (15 billion years after the Big Bang), stars formed into galaxies, while the universe continued to expand.

To illuminate the forces which constitute the preconditions immediately before the Big Bang, we have to delve into particle physics.

According to Hawking (H66), within the existing wave/particle duality not only can light and gravity be described in terms of particles, but equally so can everything else that exists in the universe. Furthermore, particles possess a property called spin. According to the quark theory, all matter consists of quarks or leptons, which are the only true elementary particles; however, no quarks have yet been identified experimentally. Hawking points out that 20 years ago it was thought that quarks were the ultimate elementary particles, meaning that they were indivisible. Now it has been found that quarks are possibly made from still smaller particles.

There now exists evidence which strengthens our belief that we are near the discovery of the ultimate substance from which the universe is made. For example, neutrinos (H185) are extremely light (possibly massless) elementary matter particles which are affected only by the weak force of gravity. On our earth, matter is mainly composed of neutrons and protons, which again consist of quarks. There is also evidence from cosmic rays that the same is true for all the matter in our galaxy. Although there is no direct evidence, it can be assumed that matter in other galaxies is also made up of quarks.

A further scrutiny of the Big Bang

Hawking (B117) states: 'At the Big Bang itself, the universe is thought to have had zero size and so to have been infinitely hot'. Strictly interpreted, the phrase 'zero size' means non-existence, while on the other hand, something that is infinitely hot must have a material existence. Consequently, Hawking's expression 'zero size' needs to be amended to read 'almost zero size'. Hawking also states that one hundred seconds after the Big Bang the temperature would have fallen to one thousand million degrees. According to him (H116), 'temperature is simply the measure of the average energy or speed of particles'.

To measure the energy embodied within the particles forming the tiny speck of matter from which the Big Bang exploded, it must be presumed that they spun at an incredibly furious pace to reach the super-temperatures claimed. The amount of heat energy available can be obtained by employing Einstein's formula, $E = me^2$. Accordingly we have to multiply the combined mass of the available particles at the time of the Big Bang with the square measure of the speed of light (which is 300,000 km per second), obtaining a figure of 90,000,000,000 (i.e. 90 billion) km per second. When multiplied with the mass of the particles available at the Big Bang (which, according to theory, ought to be equivalent to the total mass of particles in the present universe) we get the total energy available. Barnett (B55) writes: 'Einstein showed that mass and energy are equivalent and that the property called mass is simply concentrated energy. In other words, matter is energy and energy is matter and the distinction is simply one of temporary state'.

What actually happens in nature is that in every case of energy conversion a specific mass of energy-producing particles is converted into another type of matter, again producing energy. In the process of such conversions there are produced the photons of light, the gravitons of gravity, the quarks, and any or all other types of particles that populate the universe and constitute the source from which atoms and molecules, and eventually all heavenly bodies which make up the present universe (including animals and humans), have derived their existence.

All this conforms with the law of the conservation of energy, which according to Hawking (H184) is 'The law of science that states that energy (or its equivalent in mass) can be neither created nor destroyed'.

Other aspects of the COBE discovery

The COBE discovery which consolidates claims that the universe was born in a Big Bang also reveals two important aspects of time. One is that the five-point scenario previously detailed inadvertently endorses the concept of absolute time as defined but rejected by Einstein. The second is that the exclusive use of conventional calendar time in the above-mentioned scenario clashes with the doubtful assumptions of the space-time curvature syndrome advanced by Einstein and Hawking, which includes voluminal space, meaning that we have here two differing time concepts which cannot be reconciled.

The first aspect of time

As mentioned earlier the COBE discovery reveals among other things that between two immensely short time intervals of less than a split second, namely between 10^{-35} and 10^{-33} seconds after the Big Bang, the universe expanded from a size of less than an atom to the size of a grapefruit. In the subsequent expansion of the universe differing events are measured in the stated intervals of two minutes, 300,000 years, two billion years, five billion years and fifteen billion years. I repeat my previous conclusions that the time measurement employed in this process is strictly based on the regular flow of conventional calendar time, which is ultimately linked to the earth/sun relationship. In our above scenario this very time concept is retroactively employed back to the very birth of the universe to measure events lasting the tiniest part of a split second. And all this despite the fact that regular calendar time was only invented about fifteen billion years after the assumed birth of the universe. In short, we have here an example of the practical use of what Einstein (B40) had defined as 'absolute time' (a definition later endorsed by Hawking), namely *'A steady unvarying inexorable time flow, streaming from the infinite past to the infinite future'* [emphasis supplied].

The 'infinite past' reaches, of course, back to the birth of the universe, as postulated by Einstein and Hawking. It stands to reason that this definition does not allow variations in the flow of time to occur (much less allow standstill of time) anywhere in the universe during its time-measured existence.

The presumed shaping of the universe

How do Hawking and other supporters of the Big Bang theory imagine the universe was shaped during its assumed 15 billion years of existence? The view is that before the Big Bang there exists neither time, nor space nor matter nor gravity, meaning that before the birth of our universe there exists absolutely nothing. In this presumed zero scenario a tiny speck of mass appeared 15 billion years ago, and in its Big Bang explosion expanded into our universe. No indication is given of how this speck of matter was born.

In order to accommodate the curvature of space and time in the creation of the universe, Hawking compares (H137) the possible shape of the universe with the round-shaped surface of the earth. As mentioned before, Hawking (H42) proposed that the inflation of the universe can be compared with the blowing-up of a balloon with a number of spots painted on it.

With the balloon expanding, the distance between any two spots increased. But none of the spots could be considered to be the centre of expansion.

One example of this is perhaps Hubble's model of the universe (B85-86), which he estimated as having a radius measuring 35 billion light years. This distance, measured from the centre, would suggest that the universe was 35 billion years old, while a lightbeam encircling it would need over 200 billion years to return to its starting point.

However, neither Hawking's nor Hubble's model would resemble the solid body of the earth, but would merely consist of the surface enveloping a hollow interior. The model would resemble the hollow peel of a giant orange. This somehow coincides, at least graphically, with the spherical sheet postulated by Einstein and Hawking, representing curved space-time. The dilemma of how this space-time sheet of unstated thickness can accommodate all the heavenly bodies and their movements is solved by the Delphic assertion among Hawking's final conclusions (H173) 'that space and time may form a finite, four-dimensional space without singularities or boundaries, like the surface of the earth, but with more dimensions'. By postulating the possibility of more than three space-dimensions for voluminal space, a proposition which defies all bounds of human understanding and scientific logic and can therefore not be rationally contemplated, Hawking gets away with his curved space-time myth.

The above attempt to explain how the universe formed and expanded leaves several questions unresolved. For example, the Big Bang is supposed to have created not only the matter of the universe but also all its space. It is maintained that before the event neither matter, nor space, existed anywhere. As presented by Einstein and Hawking, the space of the universe is identifiable with a gravitational field, which contains all heavenly bodies, including dark matter, and forms a curved surface shaped like the earth. The process of creation is also compared with the blowing up of a slowly expanding balloon (its surface compared to a curved space-time sheet), on the surface of which constellations of stars and other matter slide slowly apart like ice-skaters.

This raises two important questions. If this curved space-time sheet is shaped like a sphere, forming an envelope of unknown thickness, it must be hollow inside, presenting a huge volume of empty space, whose presence is totally ignored. On the other hand, this assumedly empty space (not mentioned in the theory), cannot be part of the gravitational field postulated by Einstein and Hawking, who present the latter as a curved space-time sheet. Whether the theory can cope with this new and intricate problem hovering behind a curtain of space-time sheet jargon is highly questionable.

The second point to be considered is that the curved space-time sheet of unspecified thickness must be supposed to contain all the existing heavenly bodies in the universe, including of course all the galaxies. Hawking describes our own slowly rotating galaxy as measuring about one hundred thousand light years across our galaxy being merely one of some hundred thousand millions of galaxies that have been discovered by astronomers – some of them containing some hundred thousand million stars. To imagine that this colossal amount of rotating galaxies, filling a space measuring millions of light years across, can be squeezed into a flat, round-shaped, space-time sheet resembling a global envelope – whose internal space volume is condemned to total oblivion by the theory – is a proposition defying physical reality and logic. Neither does it match, in structural respects, the real-life observations of contemporary astronomy.

In his chapter 'The Expanding Universe' (H40–45), Hawking points to Alexander Friedman – a Russian physicist – who as early as 1922 indicated that our universe was expanding. Hawking, referring to Friedman, proposed three scenarios: a) an expanding universe, which would eventually collapse into itself, ending in a Big Crunch; b) an expanding universe eventually levelling out; and c) an ever-expanding universe.

What unites all three, and similar projections of an expanding universe, is that, starting with the Big Bang, all the projections under review expand in only one direction.

In nearly all descriptions of the Big Bang, the event is likened to an explosion. It began when a tiny speck of super-concentrated matter and energy exploded into a super-hot fireball which, gradually expanding, created our present universe (presumed to be about 15 billion years old.) The curious fact arising from this is that many supporters of the Big Bang, while speaking of an explosion, ignore the very dynamics of an explosion, which normally produces a globally shaped distribution of pressure, acting in all directions.

Thus, contrary to the one-directional scenario, we must suggest the more realistic proposition – namely that the Big Bang resembled most likely that of a nuclear bomb exploding in mid-air. This results in a globally distributed release of pressure heading in all direction.

What Universe?

The Big Bang theory heralding the birth of *our* universe (I am stressing the word *our* as there may be others) has been generally accepted (though only tentatively) as so far the best explanation of how it all began.

Now, four years (1996) after the COBE discovery, leading experts have begun to express doubts. Among others, they ask: why was there a Big Bang, or what may have existed before? We do not know whether our universe has siblings – are there other expanding regions well removed from what we can observe? We do not know nor understand why the fundamental constants of nature have the values they do. What is missing are experimental tests to prove what were hitherto mere assumptions.

One of the most basic assumptions concerns the term 'universe' itself. While physicists like Hawking have spoken of possible other universes besides our own, postulating additionally what they call baby-universes, they avoid trying to penetrate the veil which presumably shields other universes from our view. For example, Hawking (H46), in his chapter 'The Expanding Universe', writes '...even if there are events before the big-bang, one could not use them to determine what happened beforehand. As far as we are concerned, events before the big-bang can have no consequences, so they should not form part of a scientific model of the universe [read *our* universe]. We should therefore cut them out of the model and say that time had a beginning at the big-bang.' Such a restrictive viewpoint, coming from the pen of a peer who is considered a worthy equal to the late Albert Einstein himself, is, to say the least, surprising.

As for ourselves being mortals of a lesser standing, we can not accept that the primordial speck of matter from which our universe is reputed to have arisen (being the source of the Big Bang), came from nowhere, like a rabbit pulled from a magician's empty top hat. We must therefore assume that before our universe arose there must have been another, more ancient, universe preceding our own, from which the Big Bang scenario originated, placing the primordial speck of matter into the void.

This raises the question of how the term 'universe' should be defined. Logically considered, the term 'universe' ought to mean an all-embracing phenomenon, not allowing anything outside its bounds to exist. Yet as we see from Hawking's comment, he allows the coexistence of other universes, apart from what he calls our universe.

This leaves us no other option but to redefine the term. Curiously though, the German word 'Weltall', standing for an all-embracing universe, would do the trick. Because according to the meaning of the term, there cannot exist more than one 'Weltall'. Unfortunately the English term 'universe', having originally had the same meaning as 'Weltall', has become corrupted by the Big Bang theory which applies only to *our* universe, allowing the existence of other more ancient universes.

The only solution I can suggest is to redefine the term and speak of an 'All-embracing Universe' or a 'Super-Universe' (inclusive of all other possible universes). This super-universe ought to be eternal in time duration and infinite in extent. Here the immediate question arises: how did this super-universe come into being? It is the same problem which arises when unbelievers question the existence of an almighty creator God, by asking who created Him? While The answer in both cases is 'We can never know,' in the second case the believer is at a loss to give any answer. That at least is my humble opinion because I believe that our biologically created intellect has been specifically evolved to help the survival and continuance of our species; meaning that our intellectual powers are mere servants in this process and are certainly not designed to solve the ultimate riddles of the universe – although there is of course no harm in trying.

13

Einstein: Uncertainty and Determinism

It is generally recognised that Einstein's elaboration of the formula $E = mc^2$, and his awareness of its potential, placed him among the immortals in the pantheon of science, even if his other contributions to science were to be ignored. Jules Henri Poincaré – a French mathematician and contemporary – reputedly developed the same formula but apparently did not suspect the hidden powers locked up in the tiny atom.

Based on the simple formula $E = mc^2$, Einstein calculated that one kilogram of coal, if entirely converted into energy, would yield 25 million kilowatt hours of electricity or as much as all the power plants in the USA could generate by running steadily for two months. The formula further explains how radioactive substances like radium and uranium are able to eject particles at enormous velocities and to go on doing so for millions of years. It explains how the sun and all the stars can go on radiating light and heat for billions of years. It is however doubtful that Einstein then foresaw the development of the atom bomb and other nuclear devices which now threaten the very existence of the human species, although eventually, on his initiation, President Franklin Roosevelt set up a team to develop the atom bombs which devastated the cities of Hiroshima and Nagasaki, killing hundreds of thousands in a few seconds.

Another important aspect of Einstein's creative intelligence, though this time going against the trend of the world's physics establishment, was his conditional rejection of quantum physics and of the uncertainty principle. Einstein more than once expressed the hope (B29) that the statistical method of quantum physics would prove a temporary expedient. In a letter to physicist Max Born (4 Dec., 1926) Einstein wrote, 'Quantum mechanics is very impressive. But an inner voice tells me that it is not the real thing. The theory produces a great deal but hardly brings us closer to the secret of the old man. I am at all events convinced that He does not play dice'.

Although Einstein had gained the Nobel Prize for his contribution to the quantum theory, he held that quantum mechanics was merely a good approximation to nature, being an incomplete theory. When quantum mechanics became complete, he said, it would deal with certainties. An improved theory would also allow the certain prediction of the precise moment when a single atom would disintegrate.

Even Hawking expressed his doubts about the uncertainty principle's role in quantum mechanics (H172, 173). After defining it as implying that certain pairs of quantities, such as the position and velocity of a particle, cannot both be predicted with complete accuracy, he commented ... 'The unpredictable, in terms of position, comes in only when we try to interpret the wave in terms of positions and velocities of particles. But maybe that is our mistake: maybe there are no particle positions and velocities, but only waves. It is just that we try to fit waves to our preconceived ideas of positions and velocities. The resulting mismatch is the cause of the *apparent* unpredictability'.

I maintain that there is a certain approach which can put the uncertainty principle permanently to rest. It arises from the initial experimental mistake, in testing the theory on isolated electrons. What had been overlooked then was that, when dealing with solid bodies, any idea of uncertainty has to be abandoned. All bodies, whether organic or inorganic, like my table I write this on, or even my own person, are not only swarming with electrons but are ultimately composed of the most elementary particles of matter whether known or yet to be discovered. If we now assume that electrons would act erratically or with uncertainty within solid bodies, then not only would my table and myself disintegrate into a pile of atoms, but solid bodies could never have come into existence in the first place, neither could organic life have come into being.

Einstein's rejection of the uncertainty principle resulted from his deep insight into natural events which disallowed him to believe that nature could be intrinsically uncertain and that consequently the universe had to be one of order and harmony.

Einstein never accepted the idea that the universe was governed by chance. This, and his remark that 'God never plays dice', inevitably stamped him as a believer in causality, a concept which states that every event has a prior cause. This leads directly to the idea of determinism, adherence to which Einstein never publicly acknowledged, but which is implied in his belief in an ordered universe where chance plays no part.

The apparent weakness of the determinist idea arises from the fact that it has become too closely linked to the principle of predictability. Because while the interplay of the various laws of nature known (or yet unknown) leads to definite results (just as H_2 plus O results in water = H_2O), the monumental complexity inherent in global events is beyond the capacity of human prediction. And although short term predictions of natural events are possible to a limited extent, long term prediction of events on earth and in the universe seem to be beyond the grasp of human intellectual powers.

Scientists and philosophers have for centuries argued about the idea of determinism. For us here, the problem boils down to a confrontation of determinism and predictability. Hawking, commenting on determinism (H53, 54), mentions the French scientist the Marquis de Laplace (born 1779), who 'suggested that there should be a set of scientific laws that would allow us to predict everything that would happen in the universe ... he went further to assume that there were similar laws governing everything else, including human behaviour'. I dare to suggest that Laplace was correct insofar as determinism is concerned. But he failed to recognise that humans can never be able to unravel the monumental complexities of natural events and thus predict future results.

Thus one is compelled to accept that nature works in a deterministic way but at the same time admit that events in the universe can never be subject to human predictability. Werner Heisenberg's elaboration of the uncertainty principle is a case in point. Searching for certainties inherent in the behaviour of elementary particles (certainties which are always there for the taking), he failed to unravel the complexity inherent in their behaviour. The result was that he found himself stranded with the uncertainty principle, which offered only a partial solution.

A typical example where prediction usually stumbles is long term weather forecasting. Among the climatic factors that form weather patterns are: the volatile behaviour of atmospheric pressure, variations in temperature, changes in ozone levels, the influence of altitudes, earthquakes and volcanic eruptions, human induced pollution of the environment, deforestation, and solar influences such as sun spots and sun flares.

The interplay of all these factors, each of which rests on testable laws of nature, is so complicated that it makes long term weather prediction practically impossible. Even short term weather forecasting frequently goes astray. And yet this welter of interacting laws of nature follows strict rules and is ultimately deterministic – though in human terms the end result is unpredictable.

A case where the determinist end result points at a universe of order and harmony is the biologically induced behaviour of the genes. Gene researchers Dr Craig Venter of the Institute of Genomic Research in Gaithersburg, Maryland, and Dr Hamilton Smith of the Johns Hopkins University, Baltimore, leading a team, found that each human cell contains a copy of the human genome which has about three billion base pairs and would make a skein (closely tied coil) six feet in length if all the DNA were to be unpacked from its tightly wound coils in the cell's nucleus. The longest genomes sequenced before this were those of various viruses, which range up to about 200,000 base pairs in length.

What must be further considered in making our point is that each of the billions of the genes' base parts is again composed of uncountable numbers of electrons and their subdivision composed of even more numerous elementary particles, each of which follows a certain predetermined trajectory. What is surprising is that out of this chaotic sea of trillions (or more) of volatile factors there emerges either a human or any other kind of animal organisms, each functioning as a biological unit, replicating itself from generation to generation. The conclusion we can draw from this is that unless there was a set of determinist laws producing order within this seemingly untraceable chaos, organic life could never have emerged on our earth.

Finally I present the example of television, whose results point to an ordered and harmoniously functioning universe, based on deterministic natural laws. To my humble mind, television's achievements appear almost incredible and present perhaps a zenith in technical skill and human ingenuity.

Imagine our globe, with a diameter of about 12,700 kilometres, being enveloped by a thin atmospheric layer a mere 100 kilometres high. Now, within this atmospheric cover, imperceptible to naked human senses, the following events take place, helped by satellites. I turn on my TV set in Bangkok, Thailand, and within a split second get a live transmission of US President Bill Clinton meeting Israel's

President Shimon Peres, thousands of kilometres distant (reaching half around the globe), in Washington DC, USA. The pictures received record all the events of this meeting in every detail of colour and sound.

Furthermore, an exact replica of this transmission can also be seen on the TV set in the room next to mine, as well as simultaneously on millions of TV sets throughout the world. Also, while I am watching the Washington event, thousands of TV transmitters worldwide are sending out other programmes of various kinds simultaneously, involving sports events, concerts, plays, pop music, variety programmes, news of earthquakes, plane crashes, of revolutions and wars. At the turn of a knob on my TV set, I can switch over to any of ten different channels showing different programmes of both world and local events. All this can also be watched simultaneously in most countries of the world, without any serious interference or disturbance, assuming that the TV set is in good order. For example, one could line up a hundred TV sets next to each other, all picturing simultaneously the same programme, or any alternative programme.

How is the TV miracle accomplished? The technology involves the transmission of thousands of visible moving images, by electrical means. In the process, light waves are firstly converted into electrical impulses by a TV camera and then reconverted in picture form on the screen of a cathode-ray tube receiver. In broadcast TV the receiver consists of equipment for receiving modulated radio frequency electromagnetic radiation, representing a complete television signal, which includes sound, vision and synchronising signals. The receiver is based on the superheterodyne principle (the heterodyne effect being produced by superimposing waves of differing frequency), whereby the sound and vision signals are fed to separate intermediate frequency amplifiers, detectors, and output stages.

How these receptors can separate the millions of electrical impulses criss-crossing our atmosphere thereby creating order within a welter of different images, producing perfect reproductions in sound, colour and shape, is an almost unbelievable feat of nature brought to life by human ingenuity. The conclusion is that unless there were strict natural laws regulating the interplay of all the diverse forces swarming invisible to human perception through our atmosphere, television would never have been possible.

14

Postscript

In the preceding chapter I tried to explain the basic ideas of a determinist structured universe (inadvertently indicated by Einstein himself) where events happen in an ordered and harmonious manner. I also pointed out that in such a scenario the exact course of future events remains hidden, inaccessible and unpredictable, due to the biological limitations to which human reasoning powers are subject.

Generally, cosmologists, dealing with facts and figures, sometimes wrapped in obscure mathematical equations, are less concerned with the ultimate fate of the universe, leaving such questions for philosophers and theologists to ponder. Einstein, who reputedly was of a rather materialistic and atheistic bent, was nevertheless gripped by awe when reflecting on the mysteries of the universe.

'The most beautiful and most profound emotion we can experience,' he exclaimed to his close friend Lionel Barnett, 'is the sensation of the mystical. It is the sower of all true science. To know that what is impenetrable to us really exists, manifesting itself as the highest wisdom and the most radiant beauty which our full faculties can comprehend only in their most primitive forms – this knowledge, this feeling, is at the centre of true religiousness. My religion,' he said, 'consists of a humble admiration of the unlimitable superior spirit who reveals Himself in the slight details we are able to perceive with our frail and feeble minds. The deeply emotional conviction of the presence of a superior reasoning power, which is revealed in the incomprehensible universe, forms my idea of God. He to whom this emotion is a stranger, who can no longer wonder and stand wrapped in awe, is as good as dead.'

Struck by Einstein's last phrase, I ought to declare my demise and rush to the local newspaper to insert my obituary – because I not only fail to share Einstein's exuberance about the mysteries of the universe (in which I personally play the mere part of dust particle), but I feel no emotional response whatsoever when faced with the problems of the universe, towards which I feel aloof and indifferent, taking them in my stride just as I take eating, drinking, breathing and sleeping.

Admittedly, the physical world is full of marvels and fascinating mysteries. But it equally abounds in gruesome events. In short, we live in a world full of suffering and of evil forebodings. One needs only to recall recent events in Rwanda and Burundi, where half a million people, many of them women and children, were mutilated and killed within the space of a few weeks. And this was done without any cogent reason. Similar disasters on a smaller scale are daily reported in the press and on television. In the not so distant past, millions were starved to death under Stalin and Mao Tse Tung, not to forget Hitler's holocaust, murdering millions of Jews for the sake of racial cleansing. Cruelties and abominations, though on a smaller scale, are also recorded in the pages of the Holy Bible. Tortures and acts of terrorism are daily inflicted on innocent people by religious and ethnic fanatics all over the world. To record all the wars and disasters in human history in detail would fill the libraries of the world to overflowing.

On 10 December 1995, physicist Dr Joseph Rotblat, who participated in the development of the atom bomb (the

Manhattan Project) received the Nobel Peace Prize. In his acceptance address in Stockholm he reiterated that the nuclear danger threatening human survival is more potent now than it was during the cold war. He said that there are presently 16,000 nuclear devices in the world ready to blast off at a moment's notice, and each of them has a capacity of at least 20 times, and some even more than 100 times, the potential of the atom bomb which destroyed Hiroshima.

Our contemporary world is composed of many countries where rogue statesmen are in power. They would not hesitate to use nuclear arms if they found it expedient. In the light of this, atomic scientists in the USA have constructed a clock which dangerously ticks nearer and nearer towards 12 o'clock midnight, doomsday. It is ironic that Einstein inadvertently contributed to this situation when he developed his formula $E = mc^2$ and, as one of the results, humankind finds itself now sitting on a time bomb.

It is for such and similar reasons that I cannot agree with Einstein's expression of emotional awe when faced with the intricate mysteries of the universe which physical science offers to the investigator. My emotions are much more aroused by the music of a Beethoven or Verdi, by feelings of love and friendship, compassion and pity, than by the wonders of the physical universe.

At the same time I have to admit that the writing of this book was stimulated by an elating spirit of inquiry which, despite the arduous and painful work its writing involved, has given me a great deal of satisfaction and even exuberance. However, this does not extend to making me believe in a superior power before which I would bow my head in religious awe.

Considering all the facts, I have to conclude that we live in a self-regulating universe which, although being deterministic in its details, has no common predetermined objective. Furthermore, as I have shown, its events remain unpredictable to us humans. Charles Darwin, one of the most profound thinkers of this or any age, seems to have come to similar conclusions.

More than 100 years ago, in 1859, he wrote in *The Origin of Species*: 'My theology is in a simple muddle. I cannot look at the universe as the result of blind chance, yet I can see no evidence of beneficent design, or indeed any design of any kind in details.'